THE FUTURE IS NOW

THE FUTURE IS NOW

America Confronts the New Genetics

EDITED BY
WILLIAM KRISTOL
AND ERIC COHEN

ROWMAN & LITTLEFIELD PUBLISHERS, INC.
Lanham • Boulder • New York • Oxford

ROWMAN & LITTLEFIELD PUBLISHERS, INC.

Published in the United States of America
by Rowman & Littlefield Publishers, Inc.
A Member of the Rowman & Littlefield Publishing Group
4720 Boston Way, Lanham, Maryland 20706
www.rowmanlittlefield.com

12 Hid's Copse Road
Cumnor Hill, Oxford OX2 9JJ, England

Distributed by National Book Network

British Library Cataloguing in Publication Information Available

Library of Congress Cataloging-in-Publication Data Available

ISBN 0-7425-2195-8 (alk. paper)
ISBN 0-7425-2196-6 (pbk. : alk. paper)

Printed in the United States of America

∞™ The paper used in this publication meets the minimum requirements of American National Standard for Information Sciences—Permanence of Paper for Printed Library Materials, ANSI/NISO Z39.48-1992.

CONTENTS

NOTES AND ACKNOWLEDGMENTS

Many people and organizations assisted in the production of this book. The New America Foundation, the Phillips Foundation, the New Citizenship Project, and the Lynde and Harry Bradley Foundation all provided support—both financial and intellectual—for which the authors are grateful. Eric Brown of the New America Foundation worked tirelessly on all aspects of the book, never losing enthusiasm for the project despite the long hours, low pay, and repeat trips to Georgetown University library hunting down articles from decades past. Mary Cannon of the Bioethics Project was an invaluable guide to the current debate over human cloning and stem cells. Stephanie Ingersoll was a constant source of help and support on this project—doing anything asked of her, large and small. Leo Moniz also provided valuable research assistance.

In addition, we owe a debt to many writers and thinkers on bioethics, but two in particular deserve special mention: Leon Kass and Gilbert Meilaender, whose intellectual breadth and honesty, respect for human dignity and human excellence, and moral realism about modern progress give us direction and hope for the continuing debate over biotechnology and the human future. We would also like to thank Jed Lyons of Rowman & Littlefield for his interest in this project; Steve Wrinn, formerly of Rowman & Littlefield, for taking the book on; and Mary Carpenter and Ginger Strader of Rowman & Littlefield for sponsoring it and shepherding it skillfully from beginning to end.

A collection such as this one can never include everything worth reading on a subject. Many important articles and essays had to be left out, and a few of the selections included here have been excerpted. Where long passages or sections have been cut, the text is broken up with three asterisks (***).

Where only a small passage has been cut, we have used an ellipse. We have left out most of the footnotes, with the exception of a few explanatory ones. With these caveats, everything should appear exactly as it did originally—including British spellings, occasionally awkward phrasings, and so on. We have corrected only the most obvious typos.

Parts of the introduction and concluding essay appeared, in different form, in the following previously published articles: William Kristol, "The Future Is Now," *The Weekly Standard,* Feburary 12, 2001; Eric Cohen and William Kristol, "Should Human Cloning Be Allowed? No, It's a Moral Monstrosity," *The Wall Street Journal,* December 5, 2001; Eric Cohen and William Kristol, "Dr. West and Mr. Bin Laden," *The Weekly Standard,* December 17, 2001.

We would like to thank the following authors and publications for giving us permission to reprint their work:

Huxley, Aldous. *Brave New World*. Copyright © 1932, 1960 by Aldous Huxley. Excerpts from chapters 1, 16, and 17. Reprinted by permission of HarperCollins Publishers, Inc. and the Aldous Huxley Literary Estate.

Muller, H. J. "Social Biology and Population Improvement," *Nature* 144: 521–22 (1939). Copyright © 2002 Macmillan Magazines Ltd.

Lederberg, Joshua. "Unpredictable Variety Still Rules Human Reproduction," *The Washington Post*, Sept. 30, 1967. Reprinted with permission of author. "Experimental Genetics and Human Evolution," *The American Naturalist*, vol. 100, no. 915 (Sept.-Oct. 1966): 519–31. Reprinted with permission from the University of Chicago Press and the author.

Kass, Leon R. "Genetic Tampering," *The Washington Post*, Nov. 3, 1967. "Making Babies—The New Biology and the 'Old' Morality," *The Public Interest*, no. 26, Winter 1972: 18–56. "'Making Babies' Revisited," *The Public Interest*, no. 54, Winter 1979: 32–60. "The Wisdom of Repugnance," *The New Republic*, June 2, 1997: 17–26. "Preventing a Brave New World," *The New Republic,* May 21, 2001: 30–39. "L'Chaim and Its Limits: Why Not Immortality?" *First Things,* May 2001: 17–24. All articles and excerpts reprinted with permission of the author.

Ramsey, Paul. *Fabricated Man: The Ethics of Genetic Control.* 1970. Reprinted with permission from Yale University Press. Selections from pp. 30–31, 62–64, 80–82, 95–96, 122–23, 138–39.

Watson, James D. "Moving toward the Clonal Man," *Atlantic Monthly*, May 1971: 50–53. Reprinted with permission of the author.

Himmelfarb, Gertrude. "Two Cheers (or Maybe Just One) for Progress," May 5, 1999. Reprinted from *The Wall Street Journal*. Copyright © 1999 Dow Jones & Company, Inc. All rights reserved.

Fukuyama, Francis. "A Milestone in the Conquest of Nature," June 27, 2000. Reprinted from *The Wall Street Journal*. Copyright © 2000 Dow Jones & Company, Inc. All rights reserved. "Separating Good Biotech from Bad," May 23, 2001. Reprinted from *The Wall Street Journal*. Copyright © 2001 Dow Jones & Company, Inc. All rights reserved.

Meilaender, Gilbert. "Designing Our Descendants," *First Things*, no. 109, January 2001: 25–28. Reprinted with permission. "The Point of a Ban: Or, How to Think about Stem Cell Research," *Hastings Center Report*, vol. 31, no. 1 (January–February 2001): 9–16. Reproduced by permission. Copyright © The Hastings Center.

Wolfson, Adam. "Politics in a Brave New World," *The Public Interest*, no. 142, Winter 2001: 31–43. "Liberalism and Cloning," *The Weekly Standard*, Oct. 5, 1998: 26–29. "Biodemocracy in America," *The Public Interest*, no.146, Winter 2002. A version of this essay was presented at the Ingersoll Symposium, "The Mind and Morals of the Millennial Generation," Belmont Abbey College, Belmont, North Carolina, October 19–20, 2001. Reprinted with permission.

Tribe, Laurence H. "Second Thoughts on Cloning," originally published by *The New York Times*, Dec. 5, 1997.

Krauthammer, Charles. "Of Headless Mice . . . And Men," *Time*, Jan. 19, 1998. Copyright © 1998 Time, Inc. Reprinted with permission. "Why Pro-Lifers Are Missing the Point," *Time*, Feb. 12, 2001. Copyright © 2001 Time, Inc. Reprinted with permission. "A Nightmare of a Bill," *The Washington Post*, July 27, 2001. Copyright © 2001, The Washington Post Writers Group. Reprinted with permission.

Glassman, James K. "Who's Afraid of Human Cloning?" *The Washington Post*, Feb. 10, 1998. Reprinted with permission of author.

Bailey, Ronald. "Petri Dish Politics." Reprinted with permission from the December 1999 issue of *Reason* magazine. Copyright © 2001 by Reason Foundation, 3415 S. Sepulveda Blvd., Suite 400, Los Angeles, CA 90034. www.reason.com.

Kinsley, Michael. "Reason, Faith, and Stem Cells," *The Washington Post*, Aug. 29, 2000. Reprinted with permission of author.

Bottum, J. "The Pig-Man Cometh," *The Weekly Standard*, Oct. 23, 2000: 18–19. "Against Human Cloning," *The Weekly Standard*, May 7, 2001: 9. Reprinted with permission.

Fox, Michael J. "A Crucial Election for Medical Research," originally published by *The New York Times*, Nov. 1, 2000.

Smith, Wesley. "The Politics of Stem Cells: The Good News You Never Hear," *The Weekly Standard*, March 26, 2001: 17–20. Reprinted with permission.

Cohen, Eric. "The Politics of Cloning," *The Los Angeles Times*, June 3, 2001. "Of Missile Defense and Stem Cells," *The Weekly Standard*, July 16, 2001: 15–16. "Bush's Stem-Cell Ruling: A Missouri Compromise," *The Los Angeles Times*, Aug. 12, 2001. Reprinted with permission.

Bottum, J. and William Kristol. "For a Total Ban on Human Cloning," *The Weekly Standard*, July 2/July 9, 2001: 11–12. Reprinted with permission.

Saletan, William. "Cell Out," *Slate*, July 12, 2001. Copyright © Slate/Dist. by permission of United Feature Syndicate, Inc.

Pollitt, Katha. "It's a Bird, It's a Plane, It's . . . Superclone?" *The Nation*, July 23/30, 2001: 10. Reprinted with permission of author. "Send in the Clones," Ruth Hubbard, Stuart A. Newman, and Marcy Darnovsky. Reprinted from the October 8, 2001, issue of *The Nation* with permission.

George, Robert P. "Don't Destroy Human Life," July 30, 2001. Reprinted from *The Wall Street Journal*. Copyright © 1999 Dow Jones & Company, Inc. All rights reserved.

Cohen, Eric and William Kristol. "Cloning, Stem Cells, and Beyond," *The Weekly Standard*, Aug. 13, 2001: 22–26. Reprinted with permission.

Bush, President George W. "Stem Cell Science and the Preservation of Life," originally published by *The New York Times*, Aug. 12, 2001.

Connor, Kenneth L. "Stem Cells: Bush's Broken Promise," *The Washington Post*, Aug. 11, 2001. Reprinted with permission of author.

INTRODUCTION

THE FUTURE IS NOW

William Kristol and Eric Cohen

On July 31, 2001 the House of Representatives, after heated debate, passed the first federal legislation banning human cloning. Ten days later, on August 9, President George W. Bush delivered his first special televised address to the nation, on the issue of federal funding of embryonic stem cell research—and more broadly, the moral challenges of the biogenetic revolution. "As the genius of science extends the horizons of what we can do," the president declared, "we increasingly confront complex questions about what we should do." How we deal with these questions, he added, "may well define our age."

The rapid advance of human biotechnology also raises questions about how we arrived at this point in the first place; about the aims and guiding spirit of modern biological science; about the society—our society—that embraces, tolerates, or fatefully accepts its fruits; and about the changes in American politics that will surely accompany the coming of the bio-genetic age.

This book is about these questions. It is an effort to clarify our reasoning and our choices about the new genetics and the American future, and to understand the debate that has already begun—but is only just beginning.

Before the spring of 2001, most citizens had never heard of "embryonic stem cells": those so-called "magic cells," taken from destroyed human embryos, that the most passionate advocates claim will "end" suffering and disease," and the most passionate opponents compare to Nazi-era medicine. And while the announcement of Dolly, the first cloned mammal, in 1997 spawned a brief national debate about human cloning, the issue had largely fallen out of the public view, with federal legislation banning human cloning dying in the halls of Congress in 1998.

But by the summer of 2001, the challenges of the Brave New World had been publicly joined; the nation's interest, hopes, and fears had been awakened. Congress began a new, more substantial debate on human cloning. President Bush, after months of deliberation, made his decision about embryonic stem cells and appointed a new President's Council on Bioethics.

Then, on November 25, 2001, Advanced Cell Technology (ACT), a privately-held biotech company in Worcester, MA, announced that it had created the first known embryonic human clones. In these first experiments, the embryos died very quickly. But the company's hope is that someday they will serve as a made-to-order souce of rejection-free stem cells for those sick or dying patients who are willing to use them.

When they made their cloning announcement, the scientists at ACT said that they had no interest in creating cloned human beings. But they did not seem particularly frightened by the prospect of aiding those who do. "We didn't feel that the abuse of this technology, its potential abuses, should stop us from doing what we believe is the right thing in medicine," said Dr. Michael West, the company's president and founder. "For the sake of medicine," he added, "we need to set our fears aside."

But there is, perhaps, much that should frighten us, or at least much that should give us pause. And there are many challenges beyond the now familiar moral hazards of human cloning and embryonic stem cells. A few news items from early 2001 are suggestive:

> *"Scientists have created the first genetically modified monkey, an advance that could lead to customized primates for medical research and that brings the possibility of genetic manipulation closer than ever to humans."*
> —*Washington Post,* January 12, 2001

> *"More than 135 NIH-funded projects rely specifically on fetal tissues, and many more are believed to use those tissues incidentally. . . . Among the NIH-funded studies is one in California in which human fetal tissues have been transplanted into mice to create rodents with humanized immune systems."*
> —*Washington Post,* January 26, 2001

> *"Late last year, genetic engineering watchdog groups warned that the European Union had granted a patent in December 1999 to an Australian company for a process that would allow the creation of 'chimerical' creatures—*

human/animal hybrids. . . . This patent specifically covers the possible creation of embryos made containing both 'cells from humans and mice, sheep, pigs, cattle, goats or fish.'"
—*National Catholic Register,* January 28, 2001

"One of the leading children's hospitals in Britain illegally harvested hearts, brains, eyes and other organs from thousands of dead children without the consent of their parents, according to a government report published Tuesday."
—*Los Angeles Times,* January 31, 2001

These "advances" (or horrors) did not happen overnight. For years, we have been "progressing" step by step down a road while averting our eyes from the road's destination; or even seriously considering, in a democratic way, what that destination might look like—whether it would be desirable or dehumanizing, a new heaven or a new hell. After all, it might be said, one man's creepy research is another man's medical breakthrough; one man's dystopian world of human cloning, organ harvesting, and genetic enhancements is another's man's paradise of scientific experiments and medical progress.

The debate over the new genetics is, in the largest sense, a debate about modernity itself: about the Baconian project of mastering nature (especially man's chemical and biological nature) for the "relief of man's estate," and about the "liberation" of the postmodern self from all moral, natural, and perhaps one day biological limits. It is about whether or not men and women should remake, redesign, and prefabricate themselves and their offspring, and about whether these new genetic powers will lead, quickly or eventually, to "the abolition of man."

As C. S. Lewis put it more than a half-century ago:

> From this point of view, what we call Man's power over Nature turns out to be a power exercised by some men over other men with Nature as its instrument. . . . [The] man-moulders of the new age will be armed with the powers of an incompetent state and an irresistible scientific technique: we shall get at last a race of conditioners who really can cut out posterity in what shape they please. . . . It is not that they are bad men. They are not men at all. . . . [They] have stepped into the void. Nor are their subjects necessarily unhappy men. They are not men at all: they are artifacts. Man's final conquest has proved to be the abolition of Man.

Proponents of the new genetics will counter, of course, that such warnings are nothing more than false superstitions—the same nostalgic fear-mongering that the party of the past has always used in its effort to halt human progress. In this spirit, one member of Congress recently compared those who seek to ban human cloning to "a house of cardinals" carrying out a new inquisition; another declared that "I do not think there are any appropriate limits to human knowledge, none whatsoever."

But beyond such absolute worship of progress and the obvious replies, there are legitimate questions that those who seek to guide or limit the genetic revolution must answer: What is different about the new genetics? How is it distinct from the rest of modern medicine? In what ways would it "dehumanize" man? What would life in a Brave New World look like and would it be so bad for its inhabitants? Why should there be limits on man's ability to improve his intelligence and expand his life-span? What would it actually mean to replace humans with "post-humans"? Could we stop this next stage of "evolution" even if we wanted to?

This book is an effort to grapple with these questions.

This is not, of course, a new debate. It is an old debate with a new urgency. It is also one of those rare instances when big theoretical questions and practical political questions have so closely intersected. And so, to understand where the "advances" of today might take us, it is useful to look back at how we thought about and dealt with the advances that preceded them. And to understand the dilemmas of the future, it is important to understand the debates of the present—the arguments, the factions, the alternatives, and the rhetoric. This book aims to help us do both.

In Part I of the book, we have collected some of the classic texts—the first imaginings and blueprints, the first dreams and nightmares—of the Brave New World. We then move on to the debates of the 1960s and 1970s over the wisdom of human cloning and genetic engineering—with essays from biologist Joshua Lederberg, philosopher Leon Kass, geneticist James D. Watson, and theologian Paul Ramsey. We end with more recent considerations of the shape of American society in the new genetic age.

In Part II, we attempt to chronicle the great national debate over human cloning and stem cells, which began in 1997 with the announcement of Dolly, and culminated (so far) in the stem cell and cloning debates of 2001.

In Part III, we consider some fundamental questions about biotechnology, human nature, and American democracy—and especially the meaning of human mortality in the genetic-terrorist age.

We have sought to include the best arguments on all sides, though of course many important articles and essays had to be left out. And taken as a whole, we have aimed to be fair, though we will candidly say that our intention is to awaken readers to the moral challenges of the Brave New World, so that we might set moral limits on biological "progress" before it is too late.

William Kristol
Eric Cohen
January 2002
Washington, D.C.

PART I

Brave New World?

SECTION A

Dream or Nightmare?

There were two great dystopian novels in the twentieth century: Aldous Huxley's *Brave New World* and George Orwell's *1984*. Orwell's fears seemed more relevant for the decades immediately after the book appeared in 1949; today Huxley seems more prescient. Orwell described a world of torture chambers, state propaganda, thought police, and Big Brother televisions monitoring our every movement and every word. Huxley described a world where people become enslaved to their own desires; where "they get what they want, and they never want what they can't get"; where everything is pleasant, comfortable, orderly, and amusing. It is a world without misery, old age, suffering, or yearning.

What makes it all "work" is man's biological and chemical control over himself and his destiny—from birth to death and everything in between. Babies are no longer born to parents but produced in "hatcheries." They are pre-programmed to love their biologically pre-determined fates. Once alive, they work, have sex, go to futuristic movies called "feelies," and take "soma" when they need "a holiday from the facts." There is no religion, no literature, no families, no loves, no anxiety, no loss. Such is the price for life without hardship or tragedy—a peaceful world of dehumanized selves.

What is so haunting about the novel is that many of the technologies it described in 1932 resemble technologies now in our possession or on the horizon—from birth control, to in vitro fertilization, to Ritalin and Prozac, to human cloning, and so on. And while in Huxley's world everything is controlled by a kind of futuristic welfare state, what makes the Brave New World so "successful" is that it is, in its component parts, so appealing: health, long life, freedom from anxiety, sex without responsibility. But what Huxley realized is that "perfecting" our lesser desires means forgoing our higher ones—a Faustian bargain, but one so gradually "chosen" that those who choose it never realize it. They have programmed themselves (and their descendants) not to see and not to care.

H. J. Muller, a geneticist rather than a novelist, attempted in 1939 to lay out an actual blueprint for such a genetically improved world—where man would achieve "the ultimate genetic improvement of man" and "human mastery over those more immediate evils which are so threatening [to] our modern civilization." It would be a world where "everyone might look upon 'genius,' combined of course with stability, as his birthright." Muller's dream was Huxley's nightmare.

Included here are excerpts from Aldous Huxley's *Brave New World* and the complete text of H. J. Muller's "Geneticist's Manifesto," as it became known, signed also by 22 other scientists and researchers.

SELECTIONS FROM
BRAVE NEW WORLD

Aldous Huxley

CHAPTER 1

A squat grey building of only thirty-four stories. Over the main entrance the words, CENTRAL LONDON HATCHERY AND CONDITIONING CENTRE, and, in a shield, the World State's motto, COMMUNITY, IDENTITY, STABILITY.

The enormous room on the ground floor faced towards the north. Cold for all the summer beyond the panes, for all the tropical heat of the room itself, a harsh thin light glared through the windows, hungrily seeking some draped lay figure, some pallid shape of academic goose-flesh, but finding only the glass and nickel and bleakly shining porcelain of a laboratory. Wintriness responded to wintriness. The overalls of the workers were white, their hands gloved with a pale corpse-coloured rubber. The light was frozen, dead, a ghost. Only from the yellow barrels of the microscopes did it borrow a certain rich and living substance, lying along the polished tubes like butter, streak after luscious streak in long recession down the work tables.

"And this," said the Director opening the door, "is the Fertilizing Room."

Bent over their instruments, three hundred Fertilizers were plunged, as the Director of Hatcheries and Conditioning entered the room, in the scarcely breathing silence, the absent-minded, soliloquizing hum or whistle, of absorbed concentration. A troop of newly arrived students, very young,

Aldous Huxley (1894–1963) was a British author and essayist. His many books include *Crome Yellow*, *Point Counter Point*, *The Doors of Perception*, and, most famously, *Brave New World*. Selections from chapters 1, 16, and 17, *Brave New World*, 1932.

pink and callow, followed nervously, rather abjectly, at the Director's heels. Each of them carried a notebook, in which, whenever the great man spoke, he desperately scribbled. Straight from the horse's mouth. It was a rare privilege. The D.H.C. for Central London always made a point of personally conducting his new students round the various departments...

"I shall begin at the beginning," said the D.H.C. and the more zealous students recorded his intention in their notebooks: *Begin at the beginning.* "These," he waved his hand, "are the incubators." And opening an insulated door he showed them racks upon racks of numbered test-tubes. "The week's supply of ova. Kept," he explained, "at blood heat; whereas the male gametes," and here he opened another door, "they have to be kept at thirty-five instead of thirty-seven. Full blood heat sterilizes." Rams wrapped in theremogene beget no lambs.

Still leaning against the incubators he gave them, while the pencils scurried illegibly across the pages, a brief description of the modern fertilizing process; spoke first, of course, of its surgical introduction—"the operation undergone voluntarily for the good of Society, not to mention the fact that it carries a bonus amounting to six months' salary"; continued with some account of the technique for preserving the excised ovary alive and actively developing; passed on to a consideration of optimum temperature, salinity, viscosity; referred to the liquor in which the detached and ripened eggs were kept; and, leading his charges to the work tables, actually showed them how this liquor was drawn off from the test-tubes; how it was let out drop by drop onto the specially warmed slides of the microscopes; how the eggs which it contained were inspected for abnormalities, counted and transferred to a porous receptacle; how (and he now took them to watch the operation) this receptacle was immersed in a warm bouillon containing free-swimming spermatozoa—at a minimum concentration of one hundred thousand per cubic centimetre, he insisted; and how, after ten minutes, the container was lifted out of the liquor and its contents reexamined; how, if any of the eggs remained unfertilized, it was again immersed, and, if necessary, yet again; how the fertilized ova went back to the incubators; where the Alphas and Betas remained until definitely bottled; while the Gammas, Deltas and Epsilons were brought out again, after only thirty-six hours, to undergo Bokanovsky's Process.

"Bokanovsky's Process," repeated the Director, and the students underlined the words in their little notebooks.

One egg, one embryo, one adult—normality. But a bokanovskified egg will bud, will proliferate, will divide. From eight to ninety-six buds, and every bud will grow into a perfectly formed embryo, and every embryo into a full-sized adult. Making ninety-six human beings grow where only one grew before. Progress.

"Essentially," the D.H.C. concluded, "bokanovskification consists of a series of arrests of development. We check the normal growth and, paradoxically enough, the egg responds by budding."

Responds by budding. The pencils were busy.

He pointed. On a very slowly moving band a rack-full of test-tubes was entering a large metal box, another rack-full was emerging. Machinery faintly purred. It took eight minutes for the tubes to go through, he told them. Eight minutes of hard X-rays being about as much as an egg can stand. A few died; of the rest, the least susceptible divided into two; most put out four buds; some eight; all were returned to the incubators, where the buds began to develop; then, after two days, were suddenly chilled, chilled and checked. Two, four, eight, the buds in their turn budded; and having budded were dosed almost to death with alcohol; consequently burgeoned again and having budded—bud out of bud out of bud—were thereafter—further arrest being generally fatal—left to develop in peace. By which time the original egg was in a fair way to becoming anything from eight to ninety-six embryos—a prodigious improvement, you will agree, on nature. Identical twins—but not in piddling twos and threes as in the old viviparous days, when an egg would sometimes accidentally divide; actually by dozens, by scores at a time.

"Scores," the Director repeated and flung out his arms, as though he were distributing largesse. "Scores."

But one of the students was fool enough to ask where the advantage lay.

"My good boy!" The Director wheeled sharply round on him. "Can't you see? Can't you *see*?" He raised a hand; his expression was solemn. "Bokanovsky's Process is one of the major instruments of social stability!"

Major instruments of social stability.

Standard men and women; in uniform batches. The whole of a small factory staffed with the products of a single bokanovskified egg.

"Ninety-six identical twins working ninety-six identical machines!" The voice was almost tremulous with enthusiasm. "You really know where you are. For the first time in history." He quoted the planetary motto. "Community, Identity, Stability." Grand words. "If we could bokanovskify indefinitely the whole problem would be solved."

Solved by standard Gammas, unvarying Deltas, uniform Epsilons. Millions of identical twins. The principle of mass production at last applied to biology.

"But, alas," the Director shook his head, "we can't bokanovskify indefinitely."

Ninety-six seemed to be the limit; seventy-two a good average. From the same ovary and with gametes of the same male to manufacture as many batches of identical twins as possible—that was the best (sadly a second best) that they could do. And even that was difficult.

"For in nature it takes thirty years for two hundred eggs to reach maturity. But our business is to stabilize the population at this moment, here and now. Dribbling out twins over a quarter of a century—what would be the use of that?"

Obviously, no use at all. But Podsnap's Technique had immensely accelerated the process of ripening. They could make sure of at least a hundred and fifty mature eggs within two years. Fertilize and bokanovskify—in other words, multiply by seventy-two—and you get an average of nearly eleven thousand brothers and sisters in a hundred and fifty batches of identical twins, all within two years of the same age.

"And in exceptional cases we can make one ovary yield us over fifteen thousand adult individuals."

Beckoning to a fair-haired, ruddy young man who happened to be passing at the moment. "Mr. Foster," he called. The ruddy young man approached. "Can you tell us the record for a single ovary, Mr. Foster?"

"Sixteen thousand and twelve in this Centre," Mr. Foster replied without hesitation. He spoke very quickly, had a vivacious blue eye, and took an evident pleasure in quoting figures. "Sixteen thousand and twelve; in one hundred and eighty-nine batches of identicals. But of course they've done much better," he rattled on, "in some of the tropical Centres. Singapore has often produced over sixteen thousand five hundred; and Mombasa has actually touched the seventeen thousand mark. But then they have unfair advantages. You should see the way a negro ovary responds to pituitary! It's quite astonishing, when you're used to working with European material. Still," he added, with a laugh (but the light of combat was in his eyes and the lift of his chin was challenging), "still, we mean to beat them if we can. I'm working on a wonderful Delta-Minus ovary at this moment. Only just eighteen months old. Over twelve thousand seven hundred children already, either decanted or in embryo. And still going strong. We'll beat them yet."

"That's the spirit I like!" cried the Director, and clapped Mr. Foster on the shoulder. "Come along with us, and give these boys the benefit of your expert knowledge."

Mr. Foster smiled modestly. "With pleasure." They went.

***

"Give them a few figures, Mr. Foster," said the Director, who was tired of talking.

Mr. Foster was only too happy to give them a few figures.

Two hundred and twenty metres long, two hundred wide, ten high. He pointed upwards. Like chickens drinking, the students lifted their eyes towards the distant ceiling.

Three tiers of racks: ground floor level, first gallery, second gallery.

The spidery steel-work of gallery above gallery faded away in all directions into the dark. Near them three red ghosts were busily unloading demijohns from a moving staircase.

The escalator from the Social Predestination Room.

Each bottle could be placed on one of fifteen racks, each rack, though you couldn't see it, was a conveyor traveling at the rate of thirty-three and a third centimetres an hour. Two hundred and sixty-seven days at eight metres a day. Two thousand one hundred and thirty-six metres in all. One circuit of the cellar at ground level, one on the first gallery, half on the second, and on the two hundred and sixty-seventh morning, daylight in the Decanting Room. Independent existence—so called.

"But in the interval," Mr. Foster concluded, "we've managed to do a lot to them. Oh, a very great deal." His laugh was knowing and triumphant.

"That's the spirit I like," said the Director once more. "Let's walk around. You tell them everything, Mr. Foster."

Mr. Foster duly told them.

Told them of the growing embryo on its bed of peritoneum. Made them taste the rich blood surrogate on which it fed. Explained why it had to be stimulated with placentin and thyroxin. Told them of the *corpus luteum* extract. Showed them the jets through which at every twelfth metre from zero to 2040 it was automatically injected. Spoke of those gradually increasing doses of pituitary administered during the final ninety-six metres of their course. Described the artificial maternal circulation installed in every bottle at Metre 112; showed them the reservoir of blood-surrogate, the centrifugal pump that kept the liquid moving over the placenta and drove it through the synthetic lung and waste product filter. Referred to the embryo's troublesome tendency to anaemia, to the massive doses of hog's stomach extract and foetal foal's liver with which, in consequence, it had to be supplied.

Showed them the simple mechanism by means of which, during the last two metres out of every eight, all the embryos were simultaneously shaken into familiarity with movement. Hinted at the gravity of the so-called "trauma of decanting," and enumerated the precautions taken to minimize, by a suitable training of the bottled embryo, that dangerous shock. Told them of the test for sex carried out in the neighborhood of Metre 200. Explained the system of labelling—a T for the males, a circle for the females and for those who were destined to become freemartins a question mark, black on a white ground.

"For of course," said Mr. Foster, "in the vast majority of cases, fertility is merely a nuisance. One fertile ovary in twelve hundred—that would really be quite sufficient for our purposes. But we want to have a good choice. And of course one must always have an enormous margin of safety. So we allow as many as thirty per cent of the female embryos to develop normally. The others get a dose of male sex-hormone every twenty-four metres for the rest of the course. Result: they're decanted as freemartins—structurally quite normal (except," he had to admit, "that they *do* have the slightest tendency to grow beards), but sterile. Guaranteed sterile. Which brings us at last," continued Mr. Foster, "out of the realm of mere slavish imitation of nature into the much more interesting world of human invention."

He rubbed his hands. For of course, they didn't content themselves with merely hatching out embryos: any cow could do that.

"We also predestine and condition. We decant our babies as socialized human beings, as Alphas or Epsilons, as future sewage workers or future . . ." He was going to say "future World controllers," but correcting himself, said "future Directors of Hatcheries," instead.

The D.H.C. acknowledged the compliment with a smile.

They were passing Metre 320 on Rack 11. A young Beta-Minus mechanic was busy with screwdriver and spanner on the blood-surrogate pump of a passing bottle. The hum of the electric motor deepened by fractions of a tone as he turned the nuts. Down, down . . . A final twist, a glance at the revolution counter, and he was done. He moved two paces down the line and began the same process on the next pump.

"Reducing the number of revolutions per minute," Mr. Foster explained. "The surrogate goes round slower; therefore passes through the lung at longer intervals; therefore gives the embryo less oxygen. Nothing like oxygen-shortage for keeping an embryo below par." Again he rubbed his hands.

"But why do you want to keep the embryo below par?" asked an ingenuous student.

"Ass!" said the Director, breaking a long silence. "Hasn't it occurred to you that an Epsilon embryo must have an Epsilon environment as well as an Epsilon heredity?"

It evidently hadn't occurred to him. He was covered with confusion.

"The lower the caste," said Mr. Foster, "the shorter the oxygen." The first organ affected was the brain. After that the skeleton. At seventy per cent of normal oxygen you got dwarfs. At less than seventy eyeless monsters.

"Who are no use at all," concluded Mr. Foster.

Whereas (his voice became confidential and eager), if they could discover a technique for shortening the period of maturation what a triumph, what a benefaction to Society!

"Consider the horse."

They considered it.

Mature at six; the elephant at ten. While at thirteen a man is not yet sexually mature; and is only full-grown at twenty. Hence, of course, that fruit of delayed development, the human intelligence.

"But in Epsilons," said Mr. Foster very justly, "we don't need human intelligence."

Didn't need and didn't get it. But though the Epsilon mind was mature at ten, the Epsilon body was not fit to work till eighteen. Long years of superfluous and wasted immaturity. If the physical development could be speeded up till it was as quick, say, as a cow's, what an enormous saving to the Community!

"Enormous!" murmured the students. Mr. Foster's enthusiasm was infectious.

He became rather technical; spoke of the abnormal endocrine co-ordination which made men grow so slowly; postulated a germinal mutation to account for it. Could the effects of this germinal mutation be undone? Could the individual Epsilon embryo be made a revert, by a suitable technique, to the normality of dogs and cows? That was the problem. And it was all but solved.

Pilkington, at Mombasa, had produced individuals who were sexually mature at four and full-grown at six and a half. A scientific triumph. But socially useless. Six-year-old men and women were too stupid to do even Epsilon work. And the process was an all-or-nothing one; either you failed to modify at all, or else you modified the whole way. They were still trying to find the ideal compromise between adults of twenty and adults of six. So far without success. Mr. Foster sighed and shook his head.

Their wanderings through the crimson twilight had brought them to the neighborhood of Metre 170 on Rack 9. From this point onwards Rack 9 was enclosed and the bottles performed the remainder of their journey in a kind of tunnel, interrupted here and there by openings two or three metres wide.

"Heat conditioning," said Mr. Foster.

Hot tunnels alternated with cool tunnels. Coolness was wedded to discomfort in the form of hard X-rays. By the time they were decanted the embryos had a horror of cold. They were predestined to emigrate to the tropics, to be miners and acetate silk spinners and steel workers. Later on their minds would be made to endorse the judgment of their bodies. "We condition them to thrive on heat," concluded Mr. Foster. "Our colleagues upstairs will teach them to love it."

"And that," put in the Director sententiously, "that is the secret of happiness and virtue—liking what you've got to do. All conditioning aims at that: making people like their unescapable social destiny."

CHAPTER 16

The Savage meanwhile wandered restlessly round the room, peering with a vague superficial inquisitiveness at the books in the shelves, at the soundtrack rolls and reading machine bobbins in their numbered pigeon-holes. On the table under the window lay a massive volume bound in limp black leather-surrogate, and stamped with large golden T's. He picked it up and opened it. MY LIFE AND WORK, BY OUR FORD. The book had been published at Detroit by the Society for the Propagation of Fordian Knowledge. Idly he turned the pages, read a sentence here, a paragraph there, and had just come to the conclusion that the book didn't interest him, when the door opened, and the Resident World Controller for Western Europe walked briskly into the room.

Mustapha Mond shook hands with all three of them; but it was to the Savage that he addressed himself. "So you don't much like civilization, Mr. Savage," he said.

The Savage looked at him. He had been prepared to lie, to bluster, to remain sullenly unresponsive; but, reassured by the good-humoured intelligence of the Controller's face, he decided to tell the truth, straightforwardly. "No." He shook his head.

Bernard started and looked horrified. What would the Controller think? To be labelled as the friend of a man who said that he didn't like civilization—said it openly and, of all people, to the Controller—it was terrible. "But, John," he began. A look from Mustapha Mond reduced him to an abject silence.

"Of course," the Savage went on to admit, "there are some very nice things. All that music in the air, for instance . . ."

"Sometimes a thousand twangling instruments will hum about my ears and sometimes voices."

The Savage's face lit up with a sudden pleasure. "Have you read it too?" he asked. "I thought nobody knew about that book here, in England."

"Almost nobody. I'm one of the very few. It's prohibited, you see. But as I make the laws here, I can also break them. With impunity, Mr. Marx," he added, turning to Bernard. "Which I'm afraid you *can't* do."

Bernard sank into a yet more hopeless misery.

"But why is it prohibited?" asked the Savage. In the excitement of meeting a man who had read Shakespeare he had momentarily forgotten everything else.

The Controller shrugged his shoulders. "Because it's old; that's the chief reason. We haven't any use for old things here."

"Even when they're beautiful?"

"Particularly when they're beautiful. Beauty's attractive, and we don't want people to be attracted by old things. We want them to like the new ones."

"But the new ones are so stupid and horrible. Those plays, where there's nothing but helicopters flying about and you *feel* the people kissing." He made a grimace. "Goats and monkeys!" Only in Othello's words could he find an adequate vehicle for his contempt and hatred.

"Nice tame animals, anyhow," the Controller murmured parenthetically.

"Why don't you let them see *Othello* instead?"

"I've told you; it's old. Besides, they couldn't understand it."

Yes, that was true. He remembered how Helmholtz had laughed at *Romeo and Juliet.* "Well then," he said, after a pause, "something new that's like *Othello,* and that they could understand."

"That's what we've all been wanting to write," said Helmholtz, breaking a long silence.

"And it's what you never will write," said the Controller. "Because, if it were really like *Othello* nobody could understand it, however new it might be. And if it were new, it couldn't possibly be like *Othello.*"

"Why not?"

"Yes, why not?" Helmholtz repeated. He too was forgetting the unpleasant realities of the situation. Green with anxiety and apprehension, only Bernard remembered them; the others ignored him. "Why not?"

"Because our world is not the same as Othello's world. You can't make flivvers without steel—and you can't make tragedies without social instability.

The world's stable now. People are happy; they get what they want, and they never want what they can't get. They're well off; they're safe; they're never ill; they're not afraid of death; they're blissfully ignorant of passion and old age; they're plagued with no mothers or fathers; they've got no wives, or children, or lovers to feel strongly about; they're so conditioned that they practically can't help behaving as they ought to behave. And if anything should go wrong, there's *soma.* Which you go and chuck out of the window in the name of liberty, Mr. Savage. *Liberty!"* He laughed. "Expecting Deltas to know what liberty is! And now expecting them to understand *Othello!* My good boy!"

The Savage was silent for a little. "All the same," he insisted obstinately, *"Othello's* good, *Othello's* better than those feelies."

"Of course it is," the Controller agreed. "But that's the price we have to pay for stability. You've got to choose between happiness and what people used to call high art. We've sacrificed the high art. We have the feelies and the scent organ instead."

"But they don't mean anything."

"They mean themselves; they mean a lot of agreeable sensations to the audience."

"But they're . . . they're told by an idiot."

The Controller laughed. "You're not being very polite to your friend, Mr. Watson. One of our most distinguished Emotional Engineers . . ."

"But he's right," said Helmholtz gloomily. "Because it *is* idiotic. Writing when there's nothing to say . . ."

"Precisely. But that requires the most enormous ingenuity. You're making flivvers out of the absolute minimum of steel—works of art out of practically nothing but pure sensation."

The Savage shook his head. "It all seems to me quite horrible."

"Of course it does. Actual happiness always looks pretty squalid in comparison with the overcompensations for misery. And, of course, stability isn't nearly so spectacular as instability. And being contented has none of the glamour of a good fight against misfortune, none of the picturesqueness of a struggle with temptation, or a fatal overthrow by passion or doubt. Happiness is never grand."

***

"It's curious," he [the Controller] went on after a little pause, "to read what people in the time of Our Ford used to write about scientific progress. They seemed to have imagined that it could be allowed to go on indefinitely, regardless of everything else. Knowledge was the highest good, truth the supreme value; all the rest was secondary and subordinate. True, ideas were beginning to change even then. Our Ford himself did a great deal to shift the emphasis from truth and beauty to comfort and happiness. Mass production demanded the shift. Universal happiness keeps the wheels steadily turning; truth and beauty can't. And, of course, whenever the masses seized political power, then it was happiness rather than truth and beauty that mattered. Still, in spite of everything, unrestricted scientific research was still permitted. People still went on talking about truth and beauty as though they were the sovereign goods. Right up to the time of the Nine Years' War. *That* made them change their tune all right. What's the point of truth or beauty or knowledge when the anthrax bombs are popping all around you? That was when science first began to be controlled—after the Nine Years' War. People were ready to have even their appetites controlled then. Anything for a quiet life. We've gone on controlling ever since. It hasn't been very good for truth, of course. But it's been very good for happiness. One can't have something for nothing. Happiness has got to be paid for. You're paying for it, Mr. Watson—paying because you happen to be too much interested in beauty. I was too much interested in truth; I paid too."

CHAPTER 17

"Art, science—you seem to have paid a fairly high price for your happiness," said the Savage, when they were alone. "Anything else?"

"Well, religion, of course," replied the Controller. "There used to be something called God—before the Nine Years' War. But I was forgetting; you know all about God, I suppose."

"Well . . ." The Savage hesitated. He would have liked to say something about solitude, about night, about the mesa lying pale under the moon, about the precipice, the plunge into shadowy darkness, about death. He would have liked to speak; but there were no words. Not even in Shakespeare.

The Controller, meanwhile, had crossed to the other side of the room and was unlocking a large safe set into the wall between the bookshelves. The heavy door swung open. Rummaging in the darkness within, "It's a subject," he said, "that has always had a great interest for me." He pulled out a thick black volume. "You've never read this, for example."

The Savage took it. "*The Holy Bible, containing the Old and New Testaments,*" he read aloud from the title-page.

"Nor this." It was a small book and had lost its cover.

"The Imitation of Christ."

"Nor this." He handed out another volume.

"The Varieties of Religious Experience. By William James."

"And I've got plenty more," Mustapha Mond continued, resuming his seat. "A whole collection of pornographic old books. God in the safe and Ford on the shelves." He pointed with a laugh to his avowed library—to the shelves of books, the rack full of reading-machine bobbins and sound-track rolls.

"But if you know about God, why don't you tell them?" asked the Savage indignantly. "Why don't you give them these books about God?"

"For the same reason as we don't give them *Othello:* they're old; they're about God hundreds of years ago. Not about God now."

"But God doesn't change."

"Men do, though."

"What difference does that make?"

"All the difference in the world," said Mustapha Mond. He got up again and walked to the safe. "There was a man called Cardinal Newman," he said. "A cardinal," he exclaimed parenthetically, "was a kind of Arch-Community-Songster."

" 'I Pandulph, of fair Milan, cardinal.' I've read about them in Shakespeare."

"Of course you have. Well, as I was saying, there was a man called Cardinal Newman. Ah, here's the book." He pulled it out. "And while I'm about it I'll take this one too. It's by a man called Maine de Biran. He was a philosopher, if you know what that was."

"A man who dreams of fewer things than there are in heaven and earth," said the Savage promptly.

"Quite so. I'll read you one of the things he *did* dream of in a moment. Meanwhile, listen to what this old Arch-Community-Songster said." He opened the book at the place marked by a slip of paper and began to read. " 'We are not our own any more than what we possess is our own. We did not make ourselves, we cannot be supreme over ourselves. We are not our own masters. We are God's property. Is it not our happiness thus to view the matter? Is it any happiness or any comfort, to consider that we *are* our own? It may be thought so by the young and prosperous. These may think it a great thing to have everything, as they suppose, their own way—to depend on no one—to have to think of nothing out of sight, to be without the irksomeness of continual acknowledgment, continual prayer, continual reference of what they do to the will of another. But as time goes on, they, as all men, will find that independence was not made for man—that it is an unnatural state—will do for a while, but will not carry us on safely to the end. . .' " Mustapha Mond paused, put down the first book and, picking up the other, turned over the pages. "Take this, for example," he said, and in his deep voice once more began to read: " 'A man grows old; he feels in himself that radical sense of weakness, of listlessness, of discomfort, which accompanies the advance of age; and, feeling thus, imagines himself merely sick, lulling his fears with the notion that this distressing condition is due to some particular cause, from which, as from an illness, he hopes to recover. Vain imaginings! That sickness is old age; and a horrible disease it is. They say that it is the fear of death and of what comes after death that makes men turn to religion as they advance in years. But my own experience has given me the conviction that, quite apart from any such terrors or imaginings, the religious sentiment tends to develop as we grow older; to develop because, as the passions grow calm, as the fancy and sensibilities are less excited and less excitable, our reason becomes less troubled in its working, less obscured by the images, desires and distractions, in which it used to be absorbed; whereupon God emerges as from behind a cloud; our soul feels, sees, turns towards the source of all light; turns naturally and inevitably; for now that all that gave to the world of sensations its life and charms has begun to leak away from us, now that phenomenal existence is no more bolstered up by impressions from within or from without, we feel the need to lean on something that abides, something that will never play us false—a reality, an absolute and everlasting truth. Yes, we inevitably turn to God; for this religious sentiment is of its nature so pure, so delightful to the soul that experiences it, that it makes up to us for all our other losses.' " Mustapha Mond shut the book and leaned back in his chair. "One of the numerous things in heaven and earth that these philosophers didn't dream about was this" (he waved his hand), "us, the modern world. 'You can only be independent of God while you've got youth and prosperity; independence won't take you safely to the end.' Well, we've now got youth and prosperity right up to the end. What follows? Evidently, that we can be independent of God. 'The religious sentiment will compensate us for all our losses.' But there aren't any losses for us to compensate; religious sentiment is superfluous. And why should we go hunting for a substitute for youthful desires, when youthful desires never fail? A substitute for distractions, when we go on enjoying all the old fooleries to the very last? What need have we of repose when our minds and bodies continue to delight in activity? of consolation, when we have *soma?* of something immovable, when there is the social order?"

"Then you think there is no God?"

"No, I think there quite probably is one."

"Then why? . . ."

Mustapha Mond checked him. "But he manifests himself in different ways to different men. In premodern times he manifested himself as the being that's described in these books. Now..."

"How does he manifest himself now?" asked the Savage.

"Well, he manifests himself as an absence; as though he weren't there at all."

"That's your fault."

"Call it the fault of civilization. God isn't compatible with machinery and scientific medicine and universal happiness. You must make your choice. Our civilization has chosen machinery and medicine and happiness. That's why I have to keep these books locked up in the safe. They're smut. People would be shocked if. . ."

The Savage interrupted him. "But isn't it *natural* to feel there's a God?"

"You might as well ask if it's natural to do up one's trousers with zippers," said the Controller sarcastically. "You remind me of another of those old fellows called Bradley. He defined philosophy as the finding of bad reason for what one believes by instinct. As if one believed anything by instinct! One believes things because one has been conditioned to believe them. Finding bad reasons for what one believes for other bad reasons—that's philosophy. People believe in God because they've been conditioned to believe in God."

"But all the same," insisted the Savage, "it is natural to believe in God when you're alone—quite alone, in the night, thinking about death . . ."

"But people never are alone now," said Mustapha Mond. "We make them hate solitude; and we arrange their lives so that it's almost impossible for them ever to have it."

The Savage nodded gloomily. At Malpais he had suffered because they had shut him out from the communal activities of the pueblo, in civilized London he was suffering because he could never escape from those communal activities, never be quietly alone.

"Do you remember that bit in *King Lear*?" said the Savage at last. " 'The gods are just and of our pleasant vices make instruments to plague us; the dark and vicious place where thee he got cost him his eyes,' and Edmund answers—you remember, he's wounded, he's dying—'Thou hast spoken right; 'tis true. The wheel has come full circle; I am here.' What about that now? Doesn't there seem to be a God managing things, punishing, rewarding?"

"Well, does there?" questioned the Controller in his turn. "You can indulge in any number of pleasant vices with a freemartin and run no risks of having your eyes put out by your son's mistress. 'The wheel has come full circle; I am here.' But where would Edmund be nowadays? Sitting in a pneumatic chair, with his arm round a girl's waist, sucking away at his sex-hormone chewing-gum and looking at the feelies. The gods are just. No doubt. But their code of law is dictated, in the last resort, by the people who organize society; Providence takes its cue from men."

"Are you sure?" asked the Savage. "Are you quite sure that the Edmund in that pneumatic chair hasn't been just as heavily punished as the Edmund who's wounded and bleeding to death? The gods are just. Haven't they used his pleasant vices as an instrument to degrade him?"

"Degrade him from what position? As a happy, hard-working, goods-consuming citizen he's perfect. Of course, if you choose some other standard than ours, then perhaps you might say he was degraded. But you've got to stick to one set of postulates. You can't play Electro-magnetic Golf according to the rules of Centrifugal Bumble-puppy."

"But value dwells not in particular will," said the Savage. "It holds his estimate and dignity as well wherein 'tis precious of itself as in the prizer."

"Come, come," protested Mustapha Mond, "that's going rather far, isn't it?"

"If you allowed yourselves to think of God, you wouldn't allow yourselves to be degraded by pleasant vices. You'd have a reason for bearing things patiently, for doing things with courage. I've seen it with the Indians."

"I'm sure you have," said Mustapha Mond. "But then we aren't Indians. There isn't any need for a civilized man to bear anything that's seriously unpleasant. And as for doing things—Ford forbid that he should get the idea into his head. It would upset the whole social order if men started doing things on their own."

"What about self-denial, then? If you had a God, you'd have a reason for self-denial."

"But industrial civilization is only possible when there's no self-denial. Self-indulgence up to the very limits imposed by hygiene and economics. Otherwise the wheels stop turning."

"You'd have a reason for chastity!" said the Savage, blushing a little as he spoke the words.

"But chastity means passion, chastity means neurasthenia. And passion and neurasthenia mean instability. And instability means the end of civilization. You can't have a lasting civilization without plenty of pleasant vices."

"But God's the reason for everything noble and fine and heroic. If you had a God . . ."

"My dear young friend," said Mustapha Mond, "civilization has absolutely no need of nobility or heroism. These things are symptoms of political inefficiency. In a properly organized society like ours, nobody has any opportunities for being noble or heroic. Conditions have got to be thoroughly unstable before the occasion can arise. Where there are wars, where there are divided allegiances, where there are temptations to be resisted, objects of love to be fought for or defended—there, obviously, nobility and heroism have some sense. But there aren't any wars nowadays. The greatest care is taken to prevent you from loving any one too much. There's no such thing as a divided allegiance; you're so conditioned that you can't help doing what you ought to do. And what you ought to do is on the whole so pleasant, so many of the natural impulses are allowed free play, that there really aren't any temptations to resist. And if ever, by some unlucky chance, anything unpleasant should somehow happen, why, there's always *soma* to give you a holiday from the facts. And there's always *soma* to calm your anger, to reconcile you to your enemies, to make you patient and long-suffering. In the past you could only accomplish these things by making a great effort and after years of hard moral training. Now, you swallow two or three half-gramme tablets, and there you are. Anybody can be virtuous now. You can carry at least half your mortality about in a bottle. Christianity without tears—that's what *soma* is."

"But the tears are necessary. Don't you remember what Othello said? 'If after every tempest came such calms, may the winds blow till they have wakened death.' There's a story one of the old Indians used to tell us, about the Girl of Mátaski. The young men who wanted to marry her had to do a morning's hoeing in her garden. It seemed easy; but there were flies and mosquitoes, magic ones. Most of the young men simply couldn't stand the biting and stinging. But the one that could—he got the girl."

"Charming! But in civilized countries," said the Controller, "you can have girls without hoeing for them; and there aren't any flies or mosquitoes to sting you. We got rid of them all centuries ago."

The Savage nodded, frowning. "You got rid of them. Yes, that's just like you. Getting rid of everything unpleasant instead of learning to put up with it. Whether 'tis better in the mind to suffer the slings and arrows of outrageous fortune, or to take arms against a sea of troubles and by opposing end them . . . But you don't do either. Neither suffer nor oppose. You just abolish the slings and arrows. It's too easy."

He was suddenly silent, thinking of his mother. In her room on the thirty-seventh floor, Linda had floated in a sea of singing lights and perfumed caresses—floated away, out of space, out of time, out of the prison of her memories, her habits, her aged and bloated body. And Tomakin, ex-Director of Hatcheries and Conditioning, Tomakin was still on holiday—on holiday from humiliation and pain, in a world where he could not hear those words, that derisive laughter, could not see that hideous face, feel those moist and flabby arms round his neck, in a beautiful world . . .

"What you need," the Savage went on, "is something *with* tears for a change. Nothing costs enough here."

("Twelve and a half million dollars," Henry Foster had protested when the Savage told him that. "Twelve and a half million—that's what the new Conditioning Centre cost. Not a cent less.")

"Exposing what is mortal and unsure to all that fortune, death and danger dare, even for an eggshell. Isn't there something in that?" he asked, looking up at Mustapha Mond. "Quite apart from God—though of course God would be a reason for it. Isn't there something in living dangerously?"

"There's a great deal in it," the Controller replied. "Men and women must have their adrenals stimulated from time to time."

"What?" questioned the Savage, uncomprehending.

"It's one of the conditions of perfect health. That's why we've made the V.P.S. treatments compulsory."

"V.P.S.?"

"Violent Passion Surrogate. Regularly once a month. We flood the whole system with adrenin. It's the complete physiological equivalent of fear and rage. All the tonic effects of murdering Desdemona and being murdered by Othello, without any of the inconveniences."

"But I like the inconveniences."

"We don't," said the Controller. "We prefer to do things comfortably."

"But I don't want comfort. I want God, I want poetry, I want real danger, I want freedom, I want goodness. I want sin."

"In fact," said Mustapha Mond, "you're claiming the right to be unhappy."

"All right then," said the Savage defiantly, "I'm claiming the right to be unhappy."

"Not to mention the right to grow old and ugly and impotent; the right to have syphilis and cancer; the right to have too little to eat; the right to be lousy; the right to live in constant apprehension of what may happen tomorrow; the right to catch typhoid; the right to be tortured by unspeakable pains of every kind." There was a long silence.

"I claim them all," said the Savage at last.

Mustapha Mond shrugged his shoulders. "You're welcome," he said.

SOCIAL BIOLOGY AND POPULATION IMPROVEMENT

H. J. Muller

The question, "How could the world's population be improved most effectively genetically?" raises far broader problems than the purely biological ones, problems which the biologist unavoidably encounters as soon as he tries to get the principles of his own special field put into practice. For the effective genetic improvement of mankind is dependent upon major changes in social conditions, and correlative changes in human attitudes. In the first place, there can be no valid basis for estimating and comparing the intrinsic worth of different individuals, without economic and social conditions which provide approximately equal opportunities for all members of society instead of stratifying them from birth into classes with widely different privileges.

The second major hindrance to genetic improvement lies in the economic and political conditions which foster antagonism between different peoples, nations and 'races.' The removal of race prejudices and of the unscientific doctrine that good or bad genes are the monopoly of particular peoples or of persons with features of a given kind will not be possible, however, before the conditions which make for war and economic exploitation have been eliminated. This requires some effective sort of federation of the whole world, based on the common interests of all its peoples.

Thirdly, it cannot be expected that the raising of children will be influenced actively by considerations of the worth of future generations unless

Hermann Joseph Muller (1890–1967) was an American geneticist and winner of the Nobel Prize for medicine in 1946 for his work in radiation genetics. He also wrote widely on the role of science in society, including the prospect for mankind's genetic self-improvement. This statement was originally published in *Nature* on September 16, 1939.

parents in general have a very considerable economic security and unless they are extended such adequate economic, medical, educational and other aids in the bearing and rearing of each additional child that the having of more children does not overburden either of them. As the woman is more especially affected by child-bearing and rearing, she must be given special protection to ensure that her reproductive duties do not interfere too greatly with her opportunities to participate in the life and work of the community at large. These objects cannot be achieved unless there is an organization of production primarily for the benefit of consumer and worker, unless the conditions of employment are adapted to the needs of parents and especially of mothers, and unless dwellings, towns and community services generally are reshaped with the good of children as one of their main objectives.

A fourth prerequisite for effective genetic improvement is the legalization, the universal dissemination, and the further development through scientific investigation, of ever more efficacious means of birth control, both negative and positive, that can be put into effect at all stages of the reproductive process—as by voluntary temporary or permanent sterilization, contraception, abortion (as a third line of defense), control of fertility and of the sexual cycle, artificial insemination, etc. Along with all this the development of social consciousness and responsibility in regard to the production of children is required, and this cannot be expected to be operative unless the above-mentioned economic and social conditions for its fulfilment are present, and unless the superstitious attitude towards sex and reproduction now prevalent has been replaced by a scientific and social attitude. This will result in its being regarded as an honour and a privilege, if not a duty, for a mother, married or unmarried, or for a couple, to have the best children possible, both in respect of their upbringing and of their genetic endowment, even where the latter would mean an artificial—though always voluntary—control over the process of parenthood.

Before people in general, or the State which is supposed to represent them, can be relied upon to adopt rational policies for the guidance of their reproduction, there will have to be, fifthly, a far wider spread of knowledge of biological principles and of recognition of the truth that both environment and heredity constitute dominating and inescapable complementary factors in human wellbeing, but factors both of which are under the potential control of man and admit of unlimited but interdependent progress. Betterment of environmental conditions enhances the opportunities for genetic betterment in the ways above indicated. But it must also be understood that the effect of the bettered environment is not a direct one on the germ cells and that the Lamarckian doctrine is fallacious, according to which the children of parents who have had better opportunities for physical and mental development inherit these improvements biologically, and according to which, in consequence, the dominant classes and peoples would have become genetically superior to the underprivileged ones. The intrinsic (genetic) characteristics of any generation can be better than those of the preceding generation only as a result of some kind of *selection,* that is, by those persons of the preceding generation who had a better genetic equipment having produced more offspring, on the whole, than the rest, either through conscious choice, or as an automatic result of the way in which they lived. Under modern civilized conditions such selection is far less likely to be automatic than under primitive conditions, hence some kind of conscious guidance of selection is called for. To make this possible, however, the population must first appreciate the force of the above principles, and the social value which a wisely guided selection would have.

Sixthly, conscious selection requires, in addition, an agreed direction or directions for selection to take, and these directions cannot be social ones, that is, for the good of mankind at large, unless social motives predominate in society. This in turn implies its socialized organization. The most important genetic objectives, from a social point of view, are the improvement of those genetic characteristics which make (*a*) for health, (*b*) for the complex called intelligence, and (*c*) for those temperamental qualities which favour fellow-feeling and social behavior rather than those (to-day most esteemed by many) which make for personal 'success,' as success is usually understood at present.

A more widespread understanding of biological principles will bring with it the realization that much more than the prevention of genetic deterioration is to be sought for, and that the raising of the level of the average of the population nearly to that of the highest now existing in isolated individuals, in regard to physical well-being, intelligence and temperamental qualities, is an achievement that would—so far as purely genetic considerations are concerned—be physically possible within a comparatively small number of generations. Thus everyone might look upon 'genius,' combined of course with stability, as his birthright. As the course of evolution shows, this would represent no final stage at all, but only an earnest of still further progress in the future.

The effectiveness of such progress, however, would demand increasingly extensive and intensive research in human genetics and in the numerous fields of investigation correlated therewith. This would involve the cooperation of specialists in various branches of medicine, psychology, chemistry and, not least, the social sciences, with the improvement of the inner constitution of man himself as their central theme. The organization of the human body is marvellously intricate, and the study of its genetics is beset with special difficulties which require the prosecution of research in this field to be on a much vaster scale, as well as more exact and analytical, than hitherto contemplated. This can, however, come about when men's minds are turned from war and hate and the struggle for the elementary means of subsistence to larger aims, pursued in common.

The day when economic reconstruction will reach the stage where such human forces will be released is not yet, but it is the task of this generation to prepare for it, and all steps along the way will represent a gain, not only for the possibilities of the ultimate genetic improvement of man, to a degree seldom dreamed of hitherto, but at the same time, more directly, for human mastery over those more immediate evils which are so threatening our modern civilization.

SECTION B

MAKING LIFE IN THE LABORATORY

By the late 1960s, research in laboratory animals had progressed to the point where the prospect of human cloning and genetic engineering was no longer science fiction but futuristic science. Joshua Lederberg, a Nobel Prize–winning biologist, argued that advances in genetics point "to an impending revision of the experimental design of human evolution." He described the possibility of cloning "superior" individuals and of "mingling individual human chromosomes with other mammals" to create man-animal chimeras. He wondered what society would do about the "accidentals," the human experiments gone bad, whom he feared would so frighten the general public that "rational" genetic policy would be impossible. And he pondered the impact of such experiments on "human affairs"—which he called, with disarming distance, "an interesting exercise in social science fiction."

For Paul Ramsey, one of the leading theologians and bioethicists of the twentieth century, the prospect of genetic manipulation and control was a grave moral and philosophical problem. In his classic study, *Fabricated Man* (1970), Ramsey challenged what he called the "frivolous conscience" of many scientists and researchers, "who raise ethical problems, if they raise them at all," with no intention of abandoning the research they find so interesting and promising. The rationale for much of this research, Ramsey well-understood, was a humanitarian one—to heal the sick, to improve the gene pool, to perfect man and society. The problem, he believed, is that men are neither so perfectible nor so wise, and that "any person, or any age or society, expecting ultimate success where ultimate success is not to be reached, is peculiarly apt to devise extreme and morally illegitimate means for getting there." One chapter of Ramsey's study is a critique of Lederberg's proposals for human cloning and genetic engineering—especially the conception of man upon which those proposals were made.

Around the same time, in 1971, geneticist James D. Watson testified before Congress on the prospect of human cloning. He described the science that was taking us there, including the birth of the first cloned frog in England, and urged widespread public discussion and regulation before human cloning became a possibility—or a fact—in human beings. "This is not a decision for the scientists at all," he said. "It is a decision of the general public—do you want this or not?" He called for the creation of national and international committees to promote "wide-ranging discussion . . . at the informal as well as formal legislative level, about the manifold problems which are bound to arise if test-tube conception becomes a common occurrence."

Leon Kass, who began his career as a medical researcher and doctor, agreed. In the late 1960s, he began to consider in depth the moral and political questions raised by the prospect of creating life in the laboratory. He warned even then of the great dangers: How the prospect of "test-tube babies" threatens to confound the natural relationships of mother, father, and child; to turn procreation into manufacture; to lead to experiments on prospective children; to reduce laboratory embryos to mere resources; and to lay the scientific groundwork for a new eugenics, done not under the discredited banner of creating a "master race," but the humanitarian grounds of "eradicating suffering" and "giving the best to our children." We risk, Kass warned, taking "a major step toward making man himself another one of the man-made things."

In 1979, soon after the first baby was born through *in vitro* fertilization, Kass revisited the issue of "making babies." On balance, he argued that in vitro fertilization is an acceptable treatment for infertile couples, but that what to do with the "leftover" embryos is a complicated moral problem. Should they be used for research? Killed? Adopted by other couples? Left to die "naturally"? His discussion foreshadows the moral questions at the heart of the current debate over extracting stem cells from embryos.

Included here is a 1967 exchange between Joshua Lederberg and Leon Kass in the *Washington Post*; excerpts from Lederberg's famous 1966 article in *The American Naturalist* ("Experimental Genetics and Human Evolution"); excerpts from Paul Ramsey's 1970 study *Fabricated Man*; a slightly revised version of James D. Watson's Congressional testimony, as originally published in the *Atlantic Monthly* in May 1971; and excerpts from Leon Kass's two *Public Interest* essays, "Making Babies—The New Biology and the 'Old' Morality" (1972) and "'Makings Babies' Revisited" (1979).

UNPREDICTABLE VARIETY STILL RULES HUMAN REPRODUCTION

Joshua Lederberg

Biology and politics are bridged in many places, but few connections are more salient than that between reproduction and sex. This gives an unpredictable variety to each generation.

Every new individual gets a gambler's sample of half the genes of each of his parents. The pretensions of hereditary aristocracy are then perennially refuted by the dice throws of genetic recombination by which every egg and sperm are formed.

We do recognize some exceptional examples of vegetative reproduction among higher animals, and even man. For example, identical twins occur when the original fertilized egg undergoes one cycle of vegetative division before settling down to the business of embryonic development. A twin pair is then a familiar example of a "clone," a term borrowed from botany for a family of vegetatively related individuals, literally a set of cuttings from a given source.

Except for twin births, we must look to lower animals for most common examples of clones. An earthworm can be cut into several segments, each one capable of regenerating into an intact worm. These new animals are genetically identical and thus together constitute a clone.

Examples of parthenogenesis might be clones or not, depending on how the egg is formed and activated. Dr. M. W. Olsen's parthenogenetic turkeys

Joshua Lederberg is an American geneticist and microbiologist who received the Nobel Prize in 1958 for his work in bacterial genetics. His many interests and areas of research have included genetic engineering, human cloning, artificial intelligence, biological warfare, and the study of life outside the atmosphere. This article was originally published in the *Washington Post* on September 30, 1967.

are not clonal but result from a nuclear fusion within the egg after some genetic assortment has occurred. Thus they hatch out into male birds.

New possibilities of clonal reproduction have arisen as a by-product of experiments on nuclear changes during development of frogs and toads. Drs. Robert Briggs and Thomas J. King of Philadelphia's Institute for Cancer Research started renucleating frog eggs about 15 years ago.

After the egg was fertilized in the usual way, the fusion nucleus was removed. Lacking any genetic information, such a denucleated egg would not develop at all. However, a new nucleus could be provided by some deft micro-surgery from another cell.

Briggs and King asked whether nuclei of specialized tissue cells undergo any permanent changes in the course of embryonic development. Their first experiments laid a sound experimental groundwork.

Eggs renucleated from other eggs or from early division stages of normal embryos developed normally. However, eggs renucleated from differentiated tissues of later embryonic stages would develop only to a certain characteristic point, then stop without being able to form the full variety of normal tissues.

This suggested that tissue differentiation is associated with a selective switching off of some genes in the nucleus. However, many nuclei might have been damaged by the transplant operation, the younger cells being more rugged, and therefore more successful, donors.

Dr. J. B. Gurdon, a zoologist at Oxford, has pursued such experiments with a toad species called Xenopus. Here, renucleation often succeeds even from specialized tissue samples, like gut cells from late tadpole stages.

These renucleations can be set up in batches, each egg issuing from a different cell of the same donor tadpole. Reared to full maturity, they give a clutch of toads having the same sex and every other genetic characteristic of the donor animal, in fact a clone.

Such experiments might eventually be made to work in man, perhaps within a few years, though they are not yet reported even for the laboratory mouse. If a single nucleus from a specific tissue is not competent, we might combine eggs renucleated from several different tissues to restore the full range of developmental capability.

It is an interesting exercise in social science fiction to contemplate the changes in human affairs that might come about from the generation of a few identical twins of existing personalities. Our reactions to such a fantasy will, of course, depend on just who is immortalized in this way—but if sexual reproduction were less familiar, we might make the same comment about that.

GENETIC TAMPERING

Leon R. Kass

In his column of September 30 ("Unpredictable Variety Still Rules Human Reproduction"), Joshua Lederberg discussed recent developments in asexual genetic manipulation of laboratory animals, and raised the possibility that "such experiments might eventually be made to work in man, perhaps within a few years." Human reproduction might then become predictable and controllable. Lederberg concluded as follows:

"It is an interesting exercise in social science fiction to contemplate the changes in human affairs that might come about from the generation of a few identical twins of existing personalities. Our reactions to such a fantasy will, of course, depend on just who is immortalized in this way—but if sexual reproduction were less familiar, we might make the same comment about that."

I wish to take strong exception to the casual and cavalier tone with which Lederberg touched on the implications for human society of the scientific developments he so ably described. The possibility of genetic manipulation in man raises fundamental and enormous questions—theological, moral, political. These questions must be carefully stated, the issues clearly articulated, and the alternative policies fully and soberly considered; "interesting exercises in social science fiction" are entirely inappropriate. It is unfortunate that Dr. Lederberg is either unaware of or unwilling to discuss the moral and political problems involved; it is shocking that he chooses to speak as if

Leon R. Kass is a doctor and bioethicist and serves as chairman of the President's Council on Bioethics. His many essays and books include *Toward a More Natural Science*, *The Hungry Soul*, and *The Ethics of Human Cloning* (with James Q. Wilson)—all part of his larger project of rethinking the moral assumptions of modern life, especially those of modern science. This letter was originally published in the *Washington Post* on November 3, 1967.

these questions are trivial, and as if they are reducible to our prejudices concerning the people who might be asexually propagated. Only naiveté or "hybris" can account for such a jocular approach; neither is excusable, especially in a man of Dr. Lederberg's stature, especially in a newspaper column whose purpose is to make us wiser in matters of public policy *vis-à-vis* science.

I think at least the following questions should be discussed: (1) Are the arguments for attempting genetic manipulation in man compelling reasons? Our ability to alter human reproduction does not demonstrate that it is desirable to do so. (2) Is human will sufficient authority to advocate or to attempt to clone a man? (3) Should an independent scientist carry out such an experiment in the absence of public authorization? If not, which "public" should decide—scientists, Congress, the U.N.? (4) Who should control the genetic planning? (5) Is it not likely that, as with other technological advances, genetic technology will fall into evil hands, those of an Eichmann rather than those of a Schweitzer? (6) If the attempts to clone a man result in the "production" of a defective "product" who will or should care for "it," and what rights will "it" have? If the "offspring" is sub-human, are we to consider it *murder* to destroy it? (7) What is the distinction between "human" and "sub-human"? Does not reflection on this question suggest that the programmed reproduction of man will, in fact, de-humanize him?

The development of science and technology, once begun, often proceeds without deliberated and considered decisions. Considerations of desirability rarely govern the transition from "it can be done" to "it has been done." Biologists today are under strong obligation to raise just such questions publicly so that we may deliberate *before* the new biomedical technology is an accomplished fact, a technology whose consequences will probably dwarf those which resulted from the development of the atomic bomb.

EXPERIMENTAL GENETICS AND HUMAN EVOLUTION

Joshua Lederberg

Planning based on informed foresight is the hallmark of organized human intelligence, in every theater from the personal decisions of domestic life to school bond elections to the world industrial economy. One sphere where it is hardly ever observed is the prediction and modification of human nature. The hazards of monolithic sophistocratic rationalization of fundamental human policy should not be overlooked, and medicine is wisely dedicated to the welfare of individual patients one at a time. However, though lacking machinery for global oversight, we must still find ways to cope with the population explosion, environmental pollution, clinical experimentation, the allocation of scarce resources like kidneys (transplant or artificial), even a convention on when life begins and ends, which confounds discussion of abortion and euthanasia. Concern for the biological substratum of posterity, i.e., eugenics, is divided by the same cross-purposes. Nevertheless, whether or not he dares to advocate concrete action, every student of evolution must be intrigued by what is happening to his own species (what else matters?), and especially the new evolutionary theory needed to model a self-modifying system that makes imperfect plans for its own nature.

Repeated rediscovery notwithstanding, the eugenic controversy started in the infancy of genetic science. More recently, the integration of experimental genetics and biochemistry has provoked a new line of speculation about more powerful techniques than the gradual shift of gene frequencies by selected breeding for the modification of man. This article will first recapitu-

Selections from "Experimental Genetics and Human Evolution," *The American Naturalist*, September–October 1966.

late a widely held skepticism about the criteria for the "good man" who is the aim of eugenic policy. The strategic impasse will not deter tactical assaults, but favors those with the most obvious, short-run payoff. I will then show how this points to an impending revision of the experimental design of human evolution, based on precedents already established in other species of animals and plants.

***

There has been considerable discussion of the supposed hazard to the human gene pool from the sheltering of the tacitly "unfit" by medicine or social welfare. Not so widely understood is the futility of negative-eugenic programs: most deleterious genes are represented and maintained in the population mainly by normal (conceivably sometimes supernormal?) heterozygotes. If we attack the heterozygotes as well as overtly afflicted homozygotes, almost no human being will qualify. In addition, many well-established institutions, such as the comfort of the automobile, and of heated shelters, war, and inheritance of unearned wealth or power, are equally suspect as dysgenic. It is very difficult to see how we can reconcile any aggressive negative eugenic program with humanistic aspirations for individual self-expression and the approbation of diversity. Positive eugenic programs can be defended roughly in proportion to their ineffectiveness: applied on a really effective scale they would state the same dilemmas. At present the main hazard of these proposals is the oblique even if unintended weight they may appear to give to the enforcement of negative eugenics on outcast groups.

Genetic counseling can nevertheless play an important role within the framework of personal decision and foresight for the immediate family. It can offer grave negative cautions about inbreeding and recurrence of genetic disease; it might also encourage optimists to look for compatibility or complementarity of positive attainments as a factor in mating preference. However, the public advertisement of "superior germ plasm" (sperm banks) is open to so many distortions—like most manipulations of mass taste—that its implementation would probably run very differently from its sponsors' hopes. As in adoption proceedings, the anonymity of third parties can be set aside only at great risk to the stability of family life.

***

Embedded in molecular biology are the crucial answers to grave and basic questions about aging, the major degenerative diseases, and cancer; and it seems an easy gamble that very consequential changes in life-span and the whole pattern of life are in the offing, provided only that the momentum of existing scientific effort is sustained. Quite apart from the glimpses of the bizarre that mechanical and transplanted organs may offer, this is a general issue of the utmost importance to the fabric of human relationships; we have hardly begun to face it.

It is already a very heavy burden on the conscience of our physicians that the ebbing of life is a gradual process; that the spontaneous beating of the heart is no longer the uncontrollable axiom of human life; indeed that many a "person" could be maintained indefinitely as an organ culture if there were any motive for it. Biological science already has a great deal to say and more questions to ask about the foundations of personality and its temporal continuity, which we have not begun to apply to the disposition of our own lives. The whole issue of self-identification needs scientific reexamination before we apply infinite effort to preserve a material body, many of whose molecules are transient anyhow. Inevitably, biological knowledge weighs many human beings with personal responsibility for decisions that were once relegated to divine Providence. In mythical terms, human nature began with the eating of the fruit of the Tree of Knowledge. Curiously, Genesis correlates this with the pain of childbirth, an insight that the growth of man's brain has gone beyond the safe and comfortable. However, the expulsion from Eden only postponed our access to the Tree of Life.

***

Vegetative or clonal reproduction has a certain interest as a investigative tool in human biology, and as an indispensable basis for any systematic algenics; but other arguments suggest that there will be little delay between demonstration and use. Clonality outweighs algeny at a much earlier stage of scientific sophistication, primarily because it answers the technical specifications of the eugenicists in a way that Mendelian breeding does not. If a superior individual (and presumably then genotype) is identified, why not copy it directly, rather than suffer all the risks of recombinational disruption, including those of sex. The same solace is accorded the carrier of genetic disease: why not be sure of an exact copy of yourself rather than risk a homozygous segregant; or at worst copy your spouse and allow some degree of biological parenthood. Parental disappointment in their recombinant offspring is rather more prevalent than overt disease. Less grandiose is the assurance of sex-control; nuclear transplantation is the one method now verified.

Indeed, horticultural practice verifies that a mix of sexual and clonal reproduction makes good sense for genetic design. Leave sexual reproduction for experimental purposes; when a suitable type is ascertained, take care to maintain it by clonal propagation. The Plant Patent Act already gives legal recognition to the process, and the rights of the developer are advertised "Asexual Reproduction Forbidden."

Clonality will be available to and have significant consequences from acts of individual decision—Medawar's piecemeal social engineering—given only community acquiescence or indifference to its practice. But here this simply allows the exercise of a minority attitude, possibly long before its implications for the whole community can be understood. Most of us pretend to abhor the narcissistic motives that would impel a clonist, but he (or she) will pass just that predisposing genotype intact to the clone. Wherever and for whatever motives close endogamy has prevailed before, clonism and clonishness will prevail . . .

So long as the environment remains static, the members of the clone might congratulate themselves that they had outwitted the genetic load; and they have indeed won a short-term advantage. In the human context, it is at least debatable whether sufficient latent variability to allow for any future contingency were preserved if the population were distributed among some millions of clones. From a strictly biological standpoint, tempered clonality could allow the best of both worlds—we would at least enjoy being able to observe the experiment of discovering whether a second Einstein would outdo the first one. How to temper the process and the accompanying social frictions is another problem.

The internal properties of the clone open up new possibilities, e.g., the free exchange of organ transplants with no concern for graft rejection. More uniquely human is the diversity of brains. How much of the difficulty of intimate communication between one human and another, despite the function of common learned language, arises from the discrepancy in their genetically determined neurological hardware? Monozygotic twins are notoriously sympathetic, easily able to interpret one another's minimal gestures and brief words; I know, however, of no objective studies of their economy of communication. For further argument, I will assume that genetic identity confers neurological similarity, and that this eases communication. This has never been systematically exploited as between twins, though it might be singularly useful in stressed occupations—say a pair of astronauts, or a deep-sea diver and his pump-tender, or a surgical team. It would be relatively more important in the discourse between generations, where an older clonont would teach his infant copy. A systematic division of intellectual labor would allow efficient communicants to have something useful to say to one another.

The burden of this argument is that the cultural process poses contradictory requirements of uniformity (for communication) and heterogeneity (for innovation). We have no idea where we stand on this scale. At least in certain areas—say soldiery—it is almost certain that clones would have a self-contained advantage, partly independent of, partly accentuated by the special characteristics of the genotype which is replicated. This introverted and potentially narrow-minded advantage of a clonish group may be the chief threat to a pluralistically dedicated species.

Even when nuclear transplantation has succeeded in the mouse, there would remain formidable restraints on the way to human application, and one might even doubt the further investment of experimental effort. However several lines are likely to become active. Animal husbandry, for prize cattle and racehorses, could not ignore the opportunity, just as it bore the brunt of the enterprises of artificial insemination and oval transplantation. The dormant storage of human germ plasm as sperm will be replaced by the freezing of somatic tissues to save potential donor nuclei. Experiments on the efficacy of human nuclear transplantation will continue on a somatic basis, and these tissue clones used progressively in chimeras. Human nuclei, and individual chromosomes and genes of the karyotype, will also be recombined with cells of other animal species—these experiments now well under way in cell culture. Before long we are bound to hear of tests of the effect of dosage of the human 21st chromosome on the development of the brain of the mouse or the gorilla. Extracorporeal gestation would merely accelerate these experiments. As bizarre as they seem, they are direct translations to man of classical work in experimental cytogenetics in *Drosophila* and in many plants. They need no further advance in algeny, just a small step in cell biology.

My colleagues differ widely in their reaction to the idea that anyone could conscientiously risk the crucial experiment, the first attempt to clone a man. Perhaps this will not be attempted until gestation can be monitored closely to be sure the fetus meets expectations. The mingling of individual human chromosomes with other mammals assures a gradualistic enlargement of the field and lowers the threshold of optimism or arrogance, particularly if cloning in other mammals gives incompletely predictable results.

What are the practical aims of this discussion? It might help to redirect energies now wasted on naive eugenics and to protect the community from a misapplication of genetic policy. It may sensitize students to recognize the significance of the fruition of experiments like nuclear transplantation. Most important, it may help to provoke more critical use of the lessons of history for the direction of our future. This will need a much wider participation in these concerns. It is hard enough to approach verifiable truth in experimental work; surely much wider criticism is needed for speculations whose scientific verifiability falls in inverse proportion to their human relevance. Scientists are by no means the best qualified architects of social policy, but there are two functions no one can do for them: the apprehension and interpretation of technical challenges to expose them for political action, and forethought for the balance of scientific effort that may be needed to manage such challenges. Popular trends in scientific work towards effective responses to human needs move just as slowly as other social institutions, and good work will come only from a widespread identification of scientists with these needs.

The foundations of any policy must rest on some deliberation of purpose. One test that may appeal to skeptical scientists is to ask what they admire in the trend of human history. Few will leave out the growing richness of man's inquiry about nature, about himself and his purpose. As long as we insist that this inquiry remain open, we have a pragmatic basis for a humble appreciation of the value of innumerable different approaches to life and its questions, of respect for the dignity of human life and of individuality, and we decry the arrogance that insists on an irrevocable answer to any of these questions of value. The same humility will keep open the options for human nature until their consequences to the legacy momentarily entrusted to us are fully understood. These concerns are entirely consistent with the rigorously mechanistic formulation of life which has been the systematic basis of recent progress in biological science.

Humanistic culture rests on a definition of man which we already know to be biologically vulnerable. Nevertheless the goals of our culture rest on a credo of the sanctity of human individuality. But how do we assay for *man* to demarcate him from his isolated or scrambled tissues and organs, on one side, from experimental karyotypic hybrids on another. Pragmatically, the legal privileges of humanity will remain with objects that look enough like men to grip their consciences, and whose nurture does not cost too much. Rather than superficial appearance of face or chromosomes, a more rational criterion[†] of human identity might be the potential for communication with the species, which is the foundation on which the unique glory of man is built.

[†]On further reflection I would attack any insistence on this suggestion (which I have made before) as another example of the intellectual arrogance that I decry a few sentences before—a human foible by no means egregious.

***

Eugenics is relatively inefficacious since its reasonable aims are a necessarily slow shift in the population frequencies of favorable genes. Segregation and recombination vitiate most short-range utilities. Its proponents are therefore led to advocate not only individual attention to but the widespread adoption of its techniques, and a minority of them would seek the sanction of law to enforce the doctrine. Most geneticists would insist on a deeper knowledge of human genetics before considering statutory intrusion on personal liberties in this sphere. Meanwhile there is grave danger that the minority view will lead to a confusion of the economic and social aims of rational population policy with genocide. The defensive reaction to such a confusion could be a disastrous impediment to the adoption of family planning by just those groups whose economic and educational progress most urgently demands it.

Algeny presupposes a number of scientific advances that have yet to be perfected; and its immediate application to human biology is, probably unrealistically, discounted as purely speculative. In this paper, I infer that the path to algeny already opens up two major diversions of human evolution: clonal reproduction and introgression of genetic material from other species. Indeed, the essential features of these techniques have already been demonstrated in vertebrates, namely nuclear transplantation in amphibia, and somatic hybridization of a variety of cells in culture, including human.

Paradoxically, the issue of "subhuman" hybrids may arise first, just because of the touchiness of experimentation on obviously human material. Tissue and organ cultures and transplants are already in wide experimental or therapeutic use, but there would be widespread inhibitions about risky experiments leading to an object that could be labelled as a human or parahuman infant. However, there is enormous scientific interest in organisms whose karyotype is augmented by fragments of the human chromosome set, especially as we know so little in detail of man's biological and genetic homology with other primates. This is being and will be pushed in steps as far as biology will allow, to larger and larger proportions of human genome in intact animals, and to organ combinations and chimeras with varying proportions of human, subhuman, and hybrid tissue (note actual efforts to transplant primate organs to man).

***

These are not the most congenial subjects for friendly conversation, especially if the conversants mistake comment for advocacy. If I differ from the consensus of my colleagues it may be only in suggesting a time scale of a few years rather than decades. Indeed, we will then face two risks, (1) that our scientific position is extremely unbalanced from the standpoint of its human impact, and (2) that precedents affecting the long-term rationale of social policy will be set, not on the basis of well-debated principles, but on the accidents of the first advertised examples. The accidentals might be as capricious as the nationality, batting average, or public esteem of a clonont, the handsomeness of a parahuman progeny, the private morality of the experimenters, or public awareness that man is part of the continuum of life.

SELECTIONS FROM

FABRICATED MAN: THE ETHICS OF GENETIC CONTROL

Paul Ramsey

Men ought not to play God before they learn to be men, and after they have learned to be men they will not play God.

There are, of course, theologians who affirm that the judgment that men should not play God is only a "troglodyte and 'rear-view mirror' reaction," and not a principle that can at all clarify the meaning of responsible human decision. These theologians seek to elevate or assimilate any risk-filled, vital decision to playing a divine role. Thus they avoid asking the critical questions about the meaning of man's *creaturely* responsibilities as a man or the real role of medicine and science as a human enterprise *serving* human life.

We Americans are now familiar with the views of techno-theologians—so familiar, in fact, that many of us believe they actually are theologians and that in their writings they are using theological concepts or are doing religious ethics. They are rather to be deemed priests in an age in which the cultic praise of technology is about the only form of prophecy we know, or that can gain a hearing.

(pages 138–39)

***

We need to raise the ethical questions with a serious and not a frivolous conscience. A man of frivolous conscience announces that there are ethical

Paul Ramsey (1913–1988) was a Christian theologian and one of the founding thinkers of modern bioethics. His many books include *Basic Christian Ethics*, *War and the Christian Conscience*, *The Patient as Person*, and *Ethics at the Edges of Life*. Selections from *Fabricated Man: The Ethics of Genetic Control*, Yale University Press, 1970.

quandaries ahead that we must urgently consider before the future catches up with us. By this he often means that we need to devise a new ethics that will provide the rationalization for doing in the future what men are bound to do because of new actions and interventions science will have made possible. In contrast, a man of serious conscience means to say in raising urgent ethical questions that there may be some things that men should never do. The good things that men do can be complete only by the things they refuse to do.

(pages 122–23)

***

The Christian knows no such absolutely imperative end that would justify any means. . . . He knows that there may be a great many actions that would be wrong to put forth in this world, no matter what good consequences are expected to follow from them. . . . He will approach the question of genetic control with a category of "cruel and unusual means" that he is prohibited to employ, just as he knows there are "cruel and unusual punishments" that are not to be employed in the penal code. He will ask, What are right means? no less than he asks, What are the proper objectives? And he will know in advance that any person, or any society or age, expecting ultimate success where ultimate success is not to be reached, is peculiarly apt to devise extreme and morally illegitimate means for getting there. This, he will know, can easily be the case if men go about making themselves the lords and creators of the future race of men. He will say, of course, of any historical and future-facing action in which he is morally obliged to engage: "Only the end can justify the means" (as Dean Acheson once said of foreign policy). However, because he is not wholly engaged in future-facing action or oriented upon the future consequences with the entirety of his being, he will immediately add (as Acheson did): "This is not to say that the end justifies any means, or that some ends can justify anything." An ethics of means not derived from, or dependent upon, the objectives of action is the immediate fruit of knowing that men have another end than the receding future contains.

(pages 30–31)

***

I wish now to bring under serious scrutiny the suggestions and predictions made by Joshua Lederberg, Professor of Genetics and Biology at Stanford University, in some of his recent articles and columns. This is the proposal that *clonal* reproduction may be choice-worthy as a diversion to be introduced into human evolution, and a good way for man to control and direct the future of his species. Lederberg's notions combine somatic, psychosociological, and genetic criteria for the selection of the phenotypes to be perpetuated in reproduction with an arsenal of techniques—a more exquisite arsenal than the one from which H. J. Muller drew in choosing future possible applications of molecular biology. Lederberg seems driven to his proposals because, like Muller, he believes genetic surgery or alchemy to be a remote possibility, but also because of the fact that when genetic reconstruction or repair is accomplished, it would then take twenty years to determine whether or not the experiment had succeeded in producing a superior or even an adequate phenotype. Therefore he, like Muller, believes we should begin with the existing types whose strengths we know, and find a way to ensure their replication in greater numbers. Since the procedures he envisions using in the control of man's future seem as remote as genetic engineering (or as imminent, in this age of galloping scientific and technological progress), Lederberg's crucial argument for clonal reproduction must be that it would replicate already existing men whose stature can now be measured and esteemed.

It is difficult to tell whether or not Lederberg advocates the prospect which he has commented on in several recent articles. He seems to disavow advocacy when he concludes by saying that "these are not the most congenial subjects for friendly conversation, especially if the conversants mistake comment for advocacy." He seems to be pondering possibilities that would be worthwhile if only there were some good reason for putting them into practice, and he expressly claims simply that "it is an interesting exercise in social science fiction [later: a "fantasy"] to contemplate the changes in human affairs that might come about from the generation of a few identical twins of existing personalities." Still, on certain philosophies of the unavoidability of scientific and technological progress, whatever can be done will undoubtedly be done. To predict and even to ponder the prospects, while leaving open the question of whether the experiment *should* be tried, or having only relativistic ways of answering that question, amounts to much the same thing as espousal. A *determinism* in regard to the increasing application of medical technology combined with a radical *voluntarism* in regard to man's control of the future of his own species tends to erase the distinction between predictive comment and advocacy. Even if this is not a fair representation of Lederberg's position, there is ample espousal interspersed in his comments on the future possible uses of clonal reproduction.

(pages 62–64)

***

Lederberg begins by announcing quite correctly that "scientists are by no means the best qualified architects of social policy," and by limiting their function to the essential tasks of interpreting the technical challenges facing humanity and having the forethought to determine various scientific efforts that might be put forth to meet these challenges.

But when he comes to the questions of purpose and value that must provide the foundation of any public policy, it is still a purely scientific value that heads all the rest, and a test most likely to appeal to "skeptical scientists." It is not anything morally substantative about man's nature, himself, or human purpose, but "the growing richness of man's *inquiry about* nature, about himself, and his purpose" (italics added). That indeed is the arch-scientific value. Despite his previous announcement, Lederberg makes this the overriding value. Interspersed is some truly admirable moral rhetoric about how scientific inquiry leads to "a humble appreciation of the value of different approaches to life and its questions, of respect for the dignity of human life, and of individuality." This is plainly the "humility" that will keep open the options and answer no moral question in the area of eugenics, until the scientific accounting of our genetic options is complete. It is also one that will never find any moral limit upon the choice among options. Instead there will be only limits derived from a "rigorously mechanistic formulation" of the life sciences. This view would seem to be the reason for Lederberg's decrying "the arrogance that insists on an irrevocable answer to any of these questions of value."

On the basis of this article alone, it is evident that Lederberg has made his choice between humane civilization and a civilization based on scientific-mechanistic values alone (if the latter is a proper manner of speaking). "Humanistic culture," he writes, "rests on a definition of man which we already know to be biologically vulnerable." This is a choice made not without trouble by Lederberg the man, since he knows that "the goals of our culture rest on a credo of the sanctity of the human individual." Despite this, however, Lederberg the scientist can only pose the question that for him has but one obvious answer: "How is it possible for man to demarcate himself from his isolated or scrambled tissues and organs on one side, and from experimental karyotypic hybrids on the other?"

The entire proposal that we should clone a man and proceed with mixing chromosomes from "other spheres" with human material is, therefore, simply an extrapolation of what we should do from what we can do. There are really only technical questions to be decided. At root, therefore, the proposal denies that the mishaps constitute a crucial moral problem. The mishaps are only misadventures that may or may not awaken in a capricious public a latent humanism sufficiently powerful to stop the experiment.

Specifically concerning the mishaps, Lederberg says only that "*pragmatically,* the *legal* privileges of humanity will remain with objects that *look* enough like men to grip their *consciences,* and *whose nurture does not cost too much*" (italics added). (The reader is told later that "the handsomeness of a parahuman progeny" would be a capricious test, as indeed it is.) At this point Lederberg's own latent humanism rises to the surface of his scientific pondering of the options, and he is impelled to write: "Rather than superficial appearance of face or chromosomes, a more rational criterion of human identity might be the potential for communication within the species, which is the foundation on which the unique glory of man is built." With this statement the paragraph once ended; and it seems a minimum statement of a basis for the respect and protection of human life. But this statement was promptly corrected by a footnote in the version of this article which appeared in *The American Naturalist,* and by the addition of a sentence in the text of the article for the *Bulletin of the Atomic Scientists.* The footnote stated: "On further reflection I would attack any insistence on this suggestion (which I have made before) as another example of the intellectual arrogance that I decry a few sentences above—a human foible by no means egregious." This judgment was made a part of the text in the later revision: "Insistence on this suggestion, of course, would be an example of the intellectual arrogance decried above." Thus Lederberg corrected his announcement of a possibly human criterion for demarking man from other genetic spheres and from artificially scrambled genes.

In the case of the mishaps, therefore, we are left with only the pragmatic problem of agreeing upon the line to be drawn empirically between those that *look human* enough and those that *look* subhuman or parahuman. Given this, we could proceed to manufacture, select, and eliminate progeny. This is simply a way of saying that the mishaps do not constitute a moral problem, and that the management and control of human evolution and of the future generations of men have no intrinsic limits. The procedure has only consequences, and goals "read in" to the object-matter by the experimenter and our future possible public policy managers.

(pages 80–82)

***

It may take some temerity to oppose these grand interferences for man's self-reconstruction and control over the evolutionary future, but this is a not unreasonable position. In the present age the attempt will be made to deprive us of our wits by comparing objections to schemes of progressive genetic engineering or cloning men to earlier opposition to innoculations, blood transfusions,or the control of malaria. These things are by no means to be compared: the practice of medicine in the service of life is one thing; man's unlimited self-modification of the genetic conditions of life would be quite another matter.

Nor does it suffice to say that we are already introducing vast changes in the environment, and that these environmental changes are in fact altering mankind's genetic basis. Even if it is true that "the jet airplane has already had an incalculably greater effect on human population genetics than any conceivable program of calculated eugenics would have," that is an argument which cuts both ways. Mankind has not evidenced much wisdom in the control and redirection of his environment. It would seem unreasonable to believe that by adding to his environmental follies one or another of these grand designs for reconstructing himself, man would then show sudden increase in wisdom. If genetic policy-making were not miraculously improved over public policy-making in environmental and political matters, then access to the Tree of Life (meaning genetic management of future generations) could cause grave damage. It could cause the genetic death God once promised and by his mercy withheld so that his creature, despite having sought to lay hold of godhood, might still live and perform a limited, creaturely service of life. Then would *boundless freedom* and self-determination become *boundless destruction* in its end results, even as its methods all along included the unlimited subjugation of man to his own rational designs and designers. No man or collection of men is likely to have the wisdom to rule the future in any such way.

(pages 95–96)

MOVING TOWARD THE CLONAL MAN

James D. Watson

The notion that man might sometime soon be reproduced asexually upsets many people. The main public effect of the remarkable clonal frog produced some ten years ago in Oxford by the zoologist John Gurdon has not been awe of the elegant scientific implication of this frog's existence, but fear that a similar experiment might someday be done with human cells. Until recently, however, this foreboding has seemed more like a science fiction scenario than a real problem which the human race has to live with.

For the embryological development of man does not occur free in the placid environment of a freshwater pond, in which a frog's eggs normally turn into tadpoles and then into mature frogs. Instead, the crucial steps in human embryology always occur in the highly inaccessible womb of a human female. There the growing fetus enlarges unseen, and effectively out of range of almost any manipulation except that which is deliberately designed to abort its existence. As long as all humans develop in this manner, there is no way to take the various steps necessary to insert an adult diploid nucleus from a pre-existing human into a human egg whose maternal genetic material has previously been removed. Given the continuation of the normal processes of conception and development, the idea that we might have a world populated by people whose genetic material was identical to that of previously existing people can belong only to the domain of the novelist or

James D. Watson is a biologist and geneticist, co-discoverer (with Francis Crick) of the structure of DNA, winner of the Nobel Prize in 1962, and one of the original directors of the Human Genome Project. This article, which was originally delivered as testimony before the U.S. House of Representatives Committee on Science and Astronautics, January 28, 1971, was published in the *Atlantic Monthly* in May 1971.

moviemaker, not to that of pragmatic scientists who must think only about things which can happen.

Today, however, we must face up to the fact that the unexpectedly rapid progress of R. G. Edwards and P. S. Steptoe in working out the conditions for routine test-tube conception of human eggs means that human embryological development need no longer be a process shrouded in secrecy. It can become instead an event wide-open to a variety of experimental manipulations. Already the two scientists have developed many embryos to the eight-cell stage, and a few more into blastocysts, the stage where successful implantation into a human uterus should not be too difficult to achieve. In fact, Edwards and Steptoe hope to accomplish implantation and subsequent growth into a normal baby within the coming year.

The question naturally arises, why should any woman willingly submit to the laparoscopy operation which yields the eggs to be used in test-tube conceptions? There is clearly some danger involved every time Steptoe operates. Nonetheless, he and Edwards believe that the risks are more than counterbalanced by the fact that their research may develop methods which could make their patients able to bear children. All their patients, though having normal menstrual cycles, are infertile, many because they have blocked oviducts which prevent passage of eggs into the uterus. If so, *in vitro* growth of their eggs up to the blastocyst stage may circumvent infertility, thereby allowing normal childbirth. Moreover, since the sex of a blastocyst is easily determined by chromosomal analysis, such women would have the possibility of deciding whether to give birth to a boy or a girl.

Clearly, if Edwards and Steptoe succeed, their success will be followed up in many other places. The number of such infertile women, while small on a relative percentage basis, is likely to be large on an absolute basis. Within the United States there could be 100,000 or so women who would like a similar chance to have their own babies. At the same time, we must anticipate strong, if not hysterical, reactions from many quarters. The certainty that the ready availability of this medical technique will open up the possibility of hiring out unrelated women to carry a given baby to term is bound to outrage many people. For there is absolutely no reason why the blastocyst need be implanted in the same woman from whom the pre-ovulatory eggs were obtained. Many women with anatomical complications which prohibit successful childbearing might be strongly tempted to find a suitable surrogate. And it is easy to imagine that other women who just don't want the discomforts of pregnancy would also seek this very different form of motherhood. Of even greater concern would be the potentialities for misuse by an inhumane totalitarian government.

Some very hard decisions may soon be upon us. It is not obvious, for example, that the vague potential of abhorrent misuse should weigh more strongly than the unhappiness which thousands of married couples feel when they are unable to have their own children. Different societies are likely to view the matter differently, and it would be surprising if all should come to the same conclusion. We must, therefore, assume that techniques for the *in vitro* manipulation of human eggs are likely to become general medical practice, capable of routine performance in many major countries, within some ten to twenty years.

The situation would then be ripe for extensive efforts, either legal or illegal, at human cloning. But for such experiments to be successful, techniques would have to be developed which allow the insertion of adult diploid nuclei into human eggs which previously have had their maternal haploid nucleus removed. At first sight, this task is a very tall order since human eggs are much smaller than those of frogs, the only vertebrates that have so far been cloned. Insertion by micropipettes, the device used in the case of the frog, is always likely to damage human eggs irreversibly. Recently, however, the development of simple techniques for fusing animal cells has raised the strong possibility that further refinements of the cell-fusion method will allow the routine introduction of human diploid nuclei into enucleated human eggs. Activation of such eggs to divide to become blastocysts, followed by implantation into suitable uteri, should lead to the development of healthy fetuses, and subsequent normal-appearing babies.

The growing up to adulthood of these first clonal humans could be a very startling event, a fact already appreciated by many magazine editors, one of whom commissioned a cover with multiple copies of Ringo Starr, another of whom gave us overblown multiple likenesses of the current sex goddess, Raquel Welch. It takes little imagination to perceive that different people will have highly different fantasies, some perhaps imagining the existence of countless people with the features of Picasso or Frank Sinatra or Walt Frazier or Doris Day. And would monarchs like the Shah of Iran, knowing they might never be able to have a normal male heir, consider the possibility of having a son whose genetic constitution would be identical to their own?

Clearly, even more bizarre possibilities can be thought of, and so we might have expected that many biologists, particularly those whose work impinges upon this possibility, would seriously ponder its implication, and begin a dialogue which would educate the world's citizens and offer suggestions which our legislative bodies might consider in framing national science policies. On the whole, however, this has not happened. Though a number of scientific papers devoted to the problem of genetic engineering have casually mentioned that clonal reproduction may someday be with us, the discussion to which I am party has been so vague and devoid of meaningful time estimates as to be virtually soporific.

Does this effective silence imply a conspiracy to keep the general public unaware of a potential threat to their basic ways of life? Could it be motivated by fear that the general reaction will be a further damning of all science, thereby decreasing even more the limited money available for pure research? Or does it merely tell us that most scientists do live such an ivory-tower existence that they are capable of thinking rationally only about pure science, dismissing more practical matters as subjects for the lawyers, students, clergy, and politicians to face up to?

One or both of these possibilities may explain why more scientists have not taken cloning before the public. The main reason, I suspect, is that the prospect to most biologists still looks too remote and chancy—not worthy of immediate attention when other matters, like nuclear-weapon overproliferation and pesticide and auto-exhaust pollution, present society with immediate threats to its orderly continuation. Though scientists as a group form the most future-oriented of all professions, there are few of us who concentrate on events unlikely to become reality within the next decade or two.

To almost all the intellectually most adventurous geneticists, the seemingly distant time when cloning might first occur is more to the point than its far-reaching implication, were it to be practiced seriously. For example, Stanford's celebrated geneticist, Joshua Lederberg, among the first to talk about cloning as a practical matter, now seems bored with further talk, implying that we should channel our limited influence as public citizens to the prevention of the wide-scale, irreversible damage to our genetic material that is now occurring through increasing exposure to man-created mutagenic compounds. To him, serious talk about cloning is essentially crying wolf when a tiger is already inside the walls.

This position, however, fails to allow for what I believe will be a frenetic rush to do experimental manipulation with human eggs once they have become a readily available commodity. And that is what they will be within several years after Edwards-Steptoe methods lead to the birth of the first healthy baby by a previously infertile woman. Isolated human eggs will be found in hundreds of hospitals, and given the fact that Steptoe's laparoscopy technique frequently yields several eggs from a single woman donor, not all of the eggs so obtained, even if they could be cultured to the blastocyst stage, would ever be reimplanted into female bodies. Most of these excess eggs would likely be used for a variety of valid experimental purposes, many, for example, to perfect the Edwards-Steptoe techniques. Others could be devoted to finding methods for curing certain genetic diseases, conceivably through use of cell-fusion methods which now seem to be the correct route to cloning. The temptation to try cloning itself thus will always be close at hand.

No reason, of course, dictates that such cloning experiments need occur. Most of the medical people capable of such experimentation would probably steer clear of any step which looked as though its real purpose were to clone. But it would be short-sighted to assume that everyone would instinctively recoil from such purposes. Some people may sincerely believe the world desperately needs many copies of really exceptional people if we are to fight our way out of the ever-increasing computer-mediated complexity that makes our individual brains so frequently inadequate.

Moreover, given the widespread development of the safe clinical procedures for handling human eggs, cloning experiments would not be prohibitively expensive. They need not be restricted to the super-powers. All smaller countries now possess the resources required for eventual success. Furthermore, there need not exist the coercion of a totalitarian state to provide the surrogate mothers. There already are such widespread divergences regarding the sacredness of the act of human reproduction that the boring meaninglessness of the lives of many women would be sufficient cause for their willingness to participate in such experimentation, be it legal or illegal. Thus, if the matter proceeds in its current nondirected fashion, a human being born of clonal reproduction most likely will appear on the earth within the next twenty to fifty years, and even sooner, if some nation should actively promote the venture.

The first reaction of most people to the arrival of these asexually produced children, I suspect, would be one of despair. The nature of the bond between parents and their children, not to mention everyone's values about the individual's uniqueness, could be changed beyond recognition, and by a science which they never understood but which until recently appeared to provide more good than harm. Certainly to many people, particularly those with strong religious backgrounds, our most sensible course of action would be to de-emphasize all those forms of research which would circumvent the normal sexual reproductive process. If this step were taken, experiments on cell fusion might no longer be supported by federal funds or tax-exempt organizations. Prohibition of such research would most certainly put off the day when diploid nuclei could satisfactorily be inserted into enucleated human eggs.

Even more effective would be to take steps quickly to make illegal, or to reaffirm the illegality of, any experimental work with human embryos.

Neither of the prohibitions, however, is likely to take place. In the first place, the cell-fusion technique now offers one of the best avenues for understanding the genetic basis of cancer. Today, all over the world, cancer cells are being fused with normal cells to pinpoint those specific chromosomes responsible for given forms of cancer. In addition, fusion techniques are the basis of many genetic efforts to unravel the biochemistry of diseases like cystic fibrosis or multiple sclerosis. Any attempts now to stop such work using the argument that cloning represents a greater threat than a disease like cancer is likely to be considered irresponsible by virtually anyone able to understand the matter.

Though more people would initially go along with a prohibition of work on human embryos, many may have a change of heart when they ponder the mess which the population explosion poses. The current projections are so horrendous that responsible people are likely to consider the need for more basic embryological facts much more relevant to our self-interest than the not-very-immediate threat of a few clonal men existing some decades ahead. And the potentially militant lobby of infertile couples who see test-tube conception as their only route to the joys of raising children of their own making would carry even more weight. So, scientists like Edwards are likely to get a go-ahead signal even if, almost perversely, the immediate consequences of their "population-money"-supported research will be the production of still more babies.

Complicating any effort at effective legislative guidance is the multiplicity of places where work like Edwards' could occur, thereby making unlikely the possibility that such manipulations would have the same legal (or illegal) status throughout the world. We must assume that if Edwards and Steptoe produce a really workable method for restoring fertility, large numbers of women will search out those places where it is legal (or possible), just as now they search out places where abortions can be easily obtained.

Thus, all nations formulating policies to handle the implications of *in vitro* human embryo experimentation must realize that the problem is essentially an international one. Even if one or more countries should stop such research, their action could effectively be neutralized by the response of a neighboring country. This most disconcerting impotence also holds for the United States. If our congressional representatives, upon learning where the matter now stands, should decide that they want none of it and pass very strict laws against human embryo experimentation, their action would not seriously set back the current scientific and medical momentum which brings us close to the possibility of surrogate mothers, if not human clonal reproduction. This is because the relevant experiments are being done not in the United States, but largely in England. That is partly a matter of chance, but also a consequence of the advanced state of English cell biology, which in certain areas is far more adventurous and imaginative than its American counterpart. There is no American university which has the strength in experimental embryology that Oxford possesses.

We must not assume, however, that today the important decisions lie only before the British government. Very soon we must anticipate that a number of biologists and clinicians of other countries, sensing the potential excitement, will move into this area of science. So even if the current English effort were stifled, similar experimentation could soon begin elsewhere. Thus it appears to me most desirable that as many people as possible be informed about the new ways of human reproduction and their potential consequences, both good and bad.

This is a matter far too important to be left solely in the hands of the scientific and medical communities. The belief that surrogate mothers and clonal babies are inevitable because science always moves forward, an attitude expressed to me recently by a scientific colleague, represents a form of laissez-faire nonsense dismally reminiscent of the creed that American business, if left to itself, will solve everybody's problems. Just as the success of a corporate body in making money need not set the human condition ahead, neither does every scientific advance automatically make our lives more "meaningful." No doubt the person whose experimental skill will eventually bring forth a clonal baby will be given wide notoriety. But the child who grows up knowing that the world wants another Picasso may view his creator in a different light.

I would thus hope that over the next decade wide-reaching discussion would occur, at the informal as well as formal legislative level, about the manifold problems which are bound to arise if test-tube conception becomes a common occurrence. A blanket declaration of the worldwide illegality of human cloning might be one result of a serious effort to ask the world in which direction it wished to move. Admittedly the vast effort, required for even the most limited international arrangement, will turn off some people—those who believe the matter is of marginal importance now, and that it is a red herring designed to take our minds off our callous attitudes toward war, poverty, and racial prejudice. But if we do not think about it now, the possibility of our having a free choice will one day suddenly be gone.

MAKING BABIES

THE NEW BIOLOGY AND THE "OLD" MORALITY

Leon R. Kass

Good afternoon ladies and gentlemen. This is your pilot speaking. We are flying at an altitude of 35,000 feet and a speed of 700 miles an hour. I have two pieces of news to report, one good and one bad. The bad news is that we are lost. The good news is that we are making excellent time.

—*Author unknown*

Thoughtful men have long known that the campaign for the technological conquest of nature, conducted under the banner of modern science, would someday train its guns against the commanding officer, man himself. That day is fast approaching, if not already here. New biomedical technologies are challenging many of the formulations which have served since ancient times to define the specifically human—to demarcate human beings from the beasts on the one hand, and from the gods on the other. Birth and death, the boundaries of an individual human life, are already subject to considerable manipulation. The perfection of organ transplantation and especially of mechanical organs will make possible wholesale reconstructions of the human body. Genetic engineering, a prospect already visible on the horizon, holds forth the promise of a refined control over human capacities and powers. Finally, technologies springing from the neurological and psychological sciences (e.g., electrical and chemical stimulation of the brain) will permit the manipulation

This essay was originally published in *The Public Interest,* Winter 1972.

and alteration of the higher human functions and activities—thought, speech, memory, choice, feeling, appetite, imagination, love.

The advent of these new powers for human engineering means that some men may be destined to play God, to re-create other men in their own image. This Promethean prospect has captured the imagination of scientist and layman alike, and is being hailed in some quarters as the final solution to the miseries of the human condition. But this optimism (not to say *hybris)* has been tempered by the dim but growing recognition that the use of these new powers will raise profound and difficult moral and political questions—and precisely because the objects on which they are to operate are human beings. In this essay, I consider some of these moral and political questions in connection with one group of new technologies: the technologies for making babies.

***

Why would anyone want to provide new methods for making babies? A major reason given is that, in many instances, the "old"[†] method is not possible. Despite greatly increased abilities to diagnose and to treat the causes of infertility in recent years, some couples still remain involuntarily childless. Thus, paradoxically, while the need to limit fertility becomes ever more apparent, some scientists and physicians have taken it as their duty to satisfy the natural desire of every couple to have a child, by natural or artificial means.

Some rather large questions arise here. Physicians have a duty to treat infertility by whatever means only if patients have a right to have children by whatever means. But the "right to procreate" is an ambiguous right, and certainly not an unqualified one. Whose right is it, a woman's or a couple's? Is it a right to carry and deliver (i.e., only a woman's right) or is it a right to nurture and rear? Is it a right to have your own biological child? Even if involuntary sterilization imposed by a government would violate such a right, however defined, is it "violated" or denied by sterility not imposed from without but due to disease? Is the inability to conceive a disease? Whose disease is it? Can a couple have a disease? Does infertility demand treatment wherever found? In women over seventy? In virgin girls? In men? Can these persons claim either a natural desire or natural right to have a child, which the new technologies might or must provide them? Does infertility demand

[†]This awkward use of "old" calls attention to the subtle traps laid for us by the abuse of language. In a time and place where novelty and originality are considered cardinal virtues, and when faddishness has replaced tuberculosis as the scourge of the intellectual classes, one should vigorously resist the tendency to make things attractive simply by emphasizing their newness. Is the "old" way of beginning life *merely* old, simply traditional and conventional?

treatment by any and all available means? By artificial insemination? By *in vitro* fertilization? By extracorporeal gestation? By parthenogenesis? By cloning—i.e., "xeroxing" of existing individuals by asexual reproduction?†

Simply posing these questions suggests that both the language of rights and the language of disease could lead to great difficulties in thinking about infertility. Both point to possessions or properties of single individuals, for it is an individual who bears rights and diseases. Yet infertility is a relationship as much as a condition—a relationship between husband and wife, and between generations too. More is involved than the interests of any single individual. Ultimately, to consider infertility (or even procreation) solely from the perspective of individual rights can only undermine—in thought and in practice—the bond between childbearing and the covenant of marriage. And in a technological age, to view infertility as a "disease," one demanding treatment by physicians, automatically fosters the development and encourages the use of all the new technologies mentioned above.

A second reason given for seeking new methods for making babies is that sometimes the old method is thought to be undesirable or inadequate, primarily on eugenic grounds. A diverse—and ultimately incompatible—collection of champions are presently in bed together under this rationale: patient-centered physicians and genetic counsellors seeking to prevent the transmission of inherited diseases to prospective children of carrier parents, species-centered pessimists concerned to combat the alleged deterioration of the human gene pool, and zealous optimists eager to engineer "improvements" in the human species. The new methods called for include the growth of early embryos in the laboratory with selective destruction of those who do not pass genetic muster; directed mating with eugenically selected eggs, sperm, or both; and asexual replication of existing "superior" individuals. But serious questions can be raised with respect to these ends as well. For example, we may know which diseases we would wish not to have inflicted upon ourselves and our offspring, but are we wise enough to act upon these desires? In view of our ignorance concerning why certain genes survive in our populations, can we be sure that the eradication of genetic disease (or of any single genetic disease) is biologically a sensible goal? Might it not have unanticipated genetic consequences?

†Those who seek to submerge the distinction between *natural* and *unnatural* means would do well to ponder these questions, and reflect on what they themselves mean when they speak of "a natural desire to have children" or "a [natural human] right to have children." One cannot speak of natural desires or natural human rights or, indeed, about disease, without having some notion of "normal," "natural," and "healthy" for human beings.

The species-centered goals are even more problematic. Do we know what constitutes a deterioration or an improvement in the human gene pool? One might well argue that, at least under present conditions, the crusaders against the deterioration of the species are worried about the wrong genes. After all, how many architects of the Vietnam war have suffered from Mongolism? Who uses up more of our irreplaceable natural resources and who produces more pollution—the inmates of an institution for the retarded or the graduates of Harvard College? It is probably as indisputable as it is ignored that the world suffers more from the morally and spiritually defective than from the genetically defective. Thus, it is sad that our best minds are busy fighting our genetic shortcomings while our more serious vices are allowed to multiply unmolested.

Perhaps this is too harsh a judgment. Certainly, our genetic inheritance is entrusted to us for safekeeping and not for abuse or neglect. Perhaps a case could be made for the desirability and wisdom of certain negative or even positive eugenic goals. Still, as in the treatment of infertility, we shall also have to consider which means, if any, can be justified in the service of *any* reasonable goals.

Thirdly, there are scientific goals which themselves generate new beginnings in life. In other words, there is a limit to what can be learned about the nature and regulation of fertilization, embryonic development, or gene action from lives begun in the old, undisturbed, natural manner. This is no doubt true. But if the goal is scientific knowledge of these processes for its own sake, there is little need to develop new beginnings in *human life.* Embryological experimentation in a wide range of mammals, employing all the new technological possibilities, would yield the basic understanding. There is at present no reason to believe that the fundamental mechanisms of differentiation differ in monkeys and in man. Until extensive animal studies show otherwise, the human experiments can only be given a technological and not a purely scientific justification. (Indeed, it is the philanthropic foundations interested in finding new drugs for abortion or contraception who are supporting much of the work on the laboratory growth of human embryos. For example, the work of R. G. Edwards and his colleagues in Cambridge, England, is supported by the Ford Foundation.) These technological purposes and activities (and others, such as the use of early embryos in culture to test for mutation- or cancer-producing chemicals and drugs, or to work out techniques for genetic manipulation) may well be desirable, but they need to be so identified and distinguished from the quest for knowledge simply. Adequately to assess the desirability of any specific means, and properly to weigh alternative means, requires a clear understanding of which ends are being served.

Finally, new methods for making babies are being sought precisely because they are new and because they can be sought. While not praiseworthy reasons, these certainly are important reasons, and all-too-human ones. Drawn by the promise of fame and glory, driven by the hot breath of competitors, men do what can be done. Bio-medical scientists are no less human than anyone else. Some of them will be unable to resist the lure of immortality promised the father of the first test tube baby. Moreover, regardless of their private motives, they are encouraged to pursue the novel because of the widespread and not unjustified belief that their new findings will probably help to alleviate one form or another of human suffering. They are also encouraged by that curious new breed of techno-theologian who, after having pronounced God dead, discloses that God's dying command was that mankind should undertake its limitless, no-holds-barred self-modification, by all feasible means.

So much then for reasons why some have called for and helped to promote new beginnings for human life. But what precisely is new about these new beginnings? Such life will still come from pre-existing life; no new formation from the dust on the ground is being contemplated, nothing as new—or as old—as that first genesis of life from non-living matter is in the immediate future. What is new is nothing more radical than the divorce of the generation of new human life from human sexuality, and ultimately from the confines of the human body, a separation which began with artificial insemination and which will finish with ectogenesis, the full laboratory growth of a baby from sperm to term. What is new is that sexual intercourse will no longer be needed for generating new life. (The new technologies provide the corollary to the pill: babies without sex.) This piece of novelty leads to two others: There is a new co-progenitor, the embryologist-geneticist-physician, and there is a new home for generation, the laboratory. The mysterious and intimate processes of generation are to be moved from the darkness of the womb to the bright (fluorescent) light of the laboratory, and beyond the shadow of a single doubt.

But this movement from natural darkness to artificial light has the most profound implications. What we are considering, really, are not merely new ways of beginning individual human lives but also—and this is far more important—new ways of life and new ways of viewing life and the nature of man. Man is partly defined by his origins; to be bound up with parents, siblings, ancestors, is part of what we mean by "human." By tampering with and confounding these origins, we are involved in nothing less than creating a new conception of what it means to be human.

***

The new technologies for human engineering may well be the "transition to a wholly new path of evolution." They may, therefore, mark the end of *human* life as we and all other humans have known it. It is possible that the non-human life that may take our place will be in some sense superior—though I personally think it most unlikely, and certainly not demonstrable. In either case, we are ourselves human beings; therefore, it is proper for us to have a proprietary interest in our survival, and in our survival *as human beings.* This is a difficult enough task without having to confront the prospect of a utopian, constant remaking of our biological nature with all-powerful means but with no end in view.

***

I had earlier raised the question of whether we have sufficient wisdom to embark upon new ways for making babies, on an individual scale as well as in the mass. By now it should be clear that I believe the answer must be a resounding "No." To have developed to the point of introduction such massive powers, with so little deliberation over the desirability of their use, can hardly be regarded as evidence of wisdom. And to deny that questions of desirability, of better and worse, can be the subject of rational deliberation, to deny that rationality might dictate that there are some things that we can do which we must never do—in short, to deny the need for wisdom—can only be regarded as the height of folly.

Let us simply look at what we have done in our conquest of non-human nature. We shall find there no grounds for optimism as we now consider offers to turn our technology loose on human nature. In the absence of standards to guide and restrain the use of this awesome power, we can only dehumanize man as we have despoiled our planet. The knowledge of these standards requires a wisdom we do not possess, and what is worse, we do not even seek.

But we have an alternative. In the absence of such wisdom, we can be wise enough to know that we are not wise enough. When we lack sufficient wisdom to do, wisdom consists in not doing. Restraint, caution, abstention, delay are what this second-best (and maybe only) wisdom dictates with respect to baby manufacture, and with respect to various other forms of human engineering made possible by other new biomedical technologies. It remains for another time to discuss how to give practical effect to this conclusion: how to establish reasonable procedures for monitoring, reviewing, and regulating the new technologies; how to deal with the undesirable consequences of their proper use; how to forestall or prevent the introduction of the worst innovations; how to achieve effective international controls so that one nation's folly does not lead the world into degradation. . . .

I am aware that there are some who now suffer who will not get relief, should my view prevail. Nevertheless we must measure the cost—I do not mean the financial cost—of seeking to eradicate that suffering by any and all means. In measuring the cost, we must of course evaluate each technological step in its own terms—but we can ill afford to ignore its place in the longer journey. For, defensible step by defensible step, we could willingly walk to our own degradation. The road to Brave New World is paved with sentimentality—yes, even with love and charity. Have we enough sense to turn back?

"MAKING BABIES" REVISITED

Leon R. Kass

Seven years ago in the pages of this journal, in an article entitled "Making Babies—the New Biology and the 'Old' Morality," I explored some of the moral and political questions raised by projected new powers to intervene in the processes of human reproduction. I concluded that it would be foolish to acquire and use these powers. The questions have since been debated in "bioethical" circles and in college classrooms, and they have received intermittent attention in the popular press and in sensational novels and movies. This past year they have gained the media limelight with the Del Zio suit against Columbia University, and more especially with the birth last summer in Britain of Louise Brown, the first identified human baby born following conception in the laboratory. . . .

I was asked by the Board to discuss the ethical issues raised by the proposed research on human *in vitro* fertilization, laboratory cultures of—and experimentation with—human embryos, and the intra-uterine transfer of such embryos for the purpose of assisting human generation. In addition, I was asked to comment on the appropriateness of Federal funding of such research and on the implications of this work for the provision of health care.

***

What is the status of a fertilized human egg (i.e., a human zygote) and the embryo that develops from it? How are we to regard its being? How are we to regard it morally, i.e., how are we to behave toward it? These are, alas, all-too-familiar questions. At least analogous, if not identical, questions are central to

This essay was originally published in *The Public Interest*, Winter 1979.

the abortion controversy and are also crucial in considering whether and what sort of experimentation is properly conducted on living but aborted fetuses. Would that it were possible to say that the matter is simple and obvious, and that it has been resolved to everyone's satisfaction!

But the controversy about the morality of abortion continues to rage and divide our nation. Moreover, many who favor or do not oppose abortion do so despite the fact that they regard the pre-viable fetus as a living human organism, even if less worthy of protection than a woman's desire not to give it birth. Almost everyone senses the importance of this matter for the decision about laboratory culture of, and experimentation with, human embryos. Thus, we are obliged to take up the question of the status of the embryos, in a search for the outlines of some common ground on which many of us can stand. To the best of my knowledge, the discussion which follows is not informed by any particular sectarian or religious teaching, though it may perhaps reveal that I am a person not devoid of reverence and the capacity for awe and wonder, said by some to be the core of the "religious" sentiment.

I begin by noting that the circumstances of laboratory-grown blastocysts (i.e., 3-to-6-day-old embryos) and embryos are not identical with those of the analogous cases of (1) living fetuses facing abortion and (2) living aborted fetuses used in research. First, the fetuses whose fates are at issue in abortion are unwanted, usually the result of "accidental" conception. Here, the embryos are wanted, and deliberately created, despite certain knowledge that many of them will be destroyed or discarded.[†] Moreover, the fate of these embryos is not in conflict with the wishes, interests, or alleged rights of the pregnant women. Second, though the HEW [Health, Education, and Welfare] guidelines governing fetal research permit studies conducted on the not-at-all-viable aborted fetus, such research merely takes advantage of available "products" of abortions not themselves undertaken for the sake of the research. No one has proposed and no one would sanction the deliberate production of live fetuses to be aborted for the sake of research, even very beneficial research.[††] In contrast, we are here considering the deliberate production of embryos for the express purpose of experimentation.

[†]In the British procedures, several eggs are taken from each woman and fertilized, to increase the chance of success, but only one embryo is transferred for implantation. In Dr. [Pierre] Soupart's proposed experiments, as the embryos will be produced only for the purpose of research and not for transfer, all of them will be discarded or destroyed.

[††]A perhaps justifiable exception would be the case of a universal plague on childbirth, say because of some epidemic that fatally attacks all fetuses *in utero* at age 5 months. Faced with the prospect of the end of the race, might we not condone the deliberate institution of pregnancies to provide fetuses for research, in the hope of finding a diagnosis and remedy for this catastrophic blight?

The cases may also differ in other ways. Given the present state of the art, the largest embryo under discussion is the blastocyst, a spherical, relatively undifferentiated mass of cells, barely visible to the naked eye. In appearance it does not look human; indeed, only the most careful scrutiny by the most experienced scientist might distinguish it from similar blastocysts of other mammals. If the human zygote and blastocyst are more like the animal zygote and blastocyst than they are like the 12-week-old human fetus (which already has a humanoid appearance, differentiated organs, and electrical activity of the brain), then there would be a much-diminished ethical dilemma regarding their deliberate creation and experimental use. Needless to say, there are articulate and passionate defenders of all points of view. Let us try, however, to consider the matter afresh.

First of all, the zygote and early embryonic stages are clearly alive. They metabolize, respire, and respond to changes in the environment; they grow and divide. Second, though not yet organized into distinctive parts or organs, the blastocyst is an organic whole, self-developing, genetically unique and distinct from the egg and sperm whose union marked the beginning of its career as a discrete, unfolding being. While the egg and sperm are alive as cells, something new and alive *in a different sense* comes into being with fertilization. The truth of this is unaffected by the fact that fertilization takes time and is not an instantaneous event. For after fertilization is *complete,* there exists a new individual, with its unique genetic identity, fully potent for the self-initiated development into a mature human being, if circumstances are cooperative. Though there is some sense in which the lives of egg and sperm are continuous with the life of the new organism-to-be (or, in human terms, that the parents live on in the child or child-to-be), in the decisive sense there is a discontinuity, a new beginning, with fertilization. *After* fertilization, there is continuity of subsequent development, even if the locus of the embryo alters with implantation (or birth). Any honest biologist must be impressed by these facts, and must be inclined, at least on first glance, to the view that a human life begins at fertilization. Even Dr. Robert Edwards has apparently stumbled over this truth, perhaps inadvertently, in the remark about Louise Brown attributed to him in an article by Peter Gwynne in *Science Digest*: "The last time I saw *her, she* was just eight cells in a test-tube. *She* was beautiful *then,* and she's still beautiful *now!*" [emphasis added]

But granting that a human life begins at fertilization, and comes-to-be via a continuous process thereafter, surely—one might say—the blastocyst itself is hardly a human being. I myself would agree that a blastocyst is not, in a *full* sense, a human being—or what the current fashion calls, rather arbitrarily and without clear definition, a person. It does not look like a human being nor can it do very much of what human beings do. Yet, at the same time, I must acknowledge that the human blastocyst is (1) human in origin and (2) *potentially* a mature human being, if all goes well. This too is beyond dispute; indeed it is precisely because of its peculiarly human potentialities that people propose to study *it* rather than the embryos of other mammals. The human blastocyst, even the human blastocyst *in vitro,* is not humanly nothing; it possesses a power to become what everyone will agree is a human being.

Here it may be objected that the blastocyst *in vitro* has today no such power, because there is now no way *in vitro* to bring the blastocyst to that much later fetal stage at which it might survive on its own. There are no published reports of culture of human embryos past the blastocyst stage (though this has been reported for mice). The *in vitro* blastocyst, like the 12-week-old aborted fetus, is *in this sense* not viable (i.e., it is at a stage of maturation before the stage of possible independent existence). But if we distinguish, among the *not*-viable embryos, between the *pre*-viable and the *not-at-all* viable—on the basis that the former, though not yet viable is capable of *becoming* or *being made* viable—we note a crucial difference between the blastocyst and the 12-week abortus. Unlike an aborted fetus, the blastocyst is possibly salvageable, and hence *potentially* viable *if it is transferred to a woman for implantation.* It is not strictly true that the *in vitro* blastocyst is *necessarily* not-viable. Until proven otherwise, by embryo transfer and attempted implantation, we are right to consider the human blastocyst *in vitro* as potentially a human being and, in this respect, not fundamentally different from a blastocyst *in utero.* To put the matter more forcefully, the blastocyst *in vitro* is *more* "viable," in the sense of more salvageable, than aborted fetuses at most later stages, up to say 20 weeks.

This is not to say that such a blastocyst is therefore endowed with a so-called right to life, that failure to implant it is negligent homicide, or that experimental touchings of such blastocysts constitute assault and battery. (I myself tend to reject such claims, and indeed think that the ethical questions are not best posed in terms of "rights.") But the blastocyst is not nothing; it is *at least* potential humanity, and as such it elicits, or ought to elicit, our feelings of awe and respect. In the blastocyst, even in the zygote, we face a mysterious and awesome power, a power governed by an immanent plan that may produce an indisputably and fully human being. It deserves our respect not because it has rights or claims or sentience (which it does not have at this stage), but because of what it is, now *and* prospectively.

Let us test this provisional conclusion by considering intuitively our response to two possible fates of such zygotes, blastocysts, and early embryos.

First, should such an embryo die, will we be inclined to mourn its passing? When a woman we know miscarries, we are sad—largely for *her* loss and disappointment, but perhaps also at the premature death of a life that might have been. But we do not mourn the departed fetus, nor do we seek ritually to dispose of the remains. In this respect, we do not treat even the fetus as fully one of us.

On the other hand, we would I suppose recoil even from the thought, let alone the practice—I apologize for forcing it upon the reader—of eating such embryos, should someone discover that they would provide a great delicacy, a "human caviar." The human blastocyst would be protected by our taboo against cannibalism, which insists on the humanness of human flesh and which does not permit us to treat even the flesh of the dead as if it were mere meat. *The human embryo is not mere meat; it is not just stuff; it is not a thing.*† Because of its origin and because of its capacity, it commands a higher respect.

How much more respect? As much as for a fully developed human being? My own inclination is to say "probably not," but who can be certain? Indeed, there might be prudential and reasonable grounds for an affirmative answer, partly because the presumption of ignorance ought to err in the direction of never underestimating the basis for respect of human life, partly because so many people feel very strongly that even the blastocyst is protectably human. As a first approximation, I would analogize the early embryo *in vitro* to the early embryo *in utero* (because both are potentially viable and human). On this ground alone, *the most sensible policy is to treat the early embryo as a pre-viable fetus, with constraints imposed on early embryo research at least as great as those on fetal research.*

To some this may seem excessively scrupulous. They will argue for the importance of the absence of distinctive humanoid appearance or the absence of sentience. To be sure, we would feel more restraint in invasive procedures conducted on a five-month-old or even 12-week-old living fetus than on a blastocyst. But this added restraint on inflicting suffering on a "look-alike," feeling creature in no way denies the propriety of a prior

†Some people have suggested that the embryo be regarded like a vital organ, salvaged from a newly dead corpse, usable for transplantation or research, and that its donation by egg and sperm donors be governed by the Uniform Anatomical Gift Act, which legitimates pre-mortem consent for organ donation upon death. But though this acknowledges that embryos are not things, it is mistaken in treating embryos as mere organs, thereby overlooking that they are early stages of a *complete, whole* human being. The Uniform Anatomical Gift Act does not apply to, nor should it be stretched to cover, donation of gonads, gametes (male sperm or female eggs), or—especially—zygotes and embryos.

restraint, grounded in respect for individuated, living, potential humanity. Before I would be persuaded to treat early embryos differently from later ones, I would insist on the establishment of a reasonably clear, naturally grounded boundary that separates "early" and "late," and which provides the basis for respecting "the early" less than "the late." This burden *must* be accepted by proponents of experimentation with human embryos *in vitro* if a decision to permit creating embryos for such experimentation is to be treated as ethically responsible.

***

Where does the above analysis lead in thinking about treatment of *in vitro* human embryos? I shall indicate, very briefly, the lines toward a possible policy, though that is not my major intent.

The *in vitro* fertilized embryo has four possible fates: (1) *implantation,* in the hope of producing from it a child; (2) *death,* by active "killing" or disaggregation, or by a "natural" demise; (3) use in *manipulative experimentation*—embryological, genetic, etc.; (4) use in attempts at *perpetuation in vitro* beyond the blastocyst stage, ultimately, perhaps, to viability. I will not now consider this fourth and future possibility, though I would suggest that full laboratory growth of an embryo into a viable human being (i.e., ectogenesis), while perfectly compatible with respect owed to its potential humanity as an individual, may be incompatible with the kind of respect owed to its humanity that is grounded in the bonds of lineage and the nature of parenthood.

On the strength of my analysis of the status of the embryo, and the respect due it, no objection would be raised to implantation. *In vitro* fertilization and embryo transfer to treat infertility, as in the case of Mr. and Mrs. Brown, is perfectly compatible with a respect and reverence for human life, including potential human life. Moreover, no disrespect is intended or practiced by the mere fact that several eggs are removed for fertilization, to increase the chance of success. Were it possible to guarantee successful fertilization and normal growth with a single egg, no more would need to be obtained. Assuming nothing further is done with the unimplanted embryos, there is nothing disrespectful going on. The demise of the unimplanted embryos would be analogous to the loss of numerous embryos wasted in the normal *in vivo* attempts to generate a child. It is estimated that over 50 percent of eggs successfully fertilized during unprotected sexual intercourse fail to implant, or do not remain implanted, in the uterine wall, and are shed soon thereafter, before a diagnosis of pregnancy could be made. Any couple attempting to conceive a child tacitly accepts such embryonic wastage as the perfectly acceptable price to be paid for the birth of a (usually) healthy child. Current procedures to initiate pregnancy with laboratory fertilization thus differ from the natural "procedure" in that what would normally be spread over four or five months *in vivo* is compressed into a single effort, using all at once a four or five months' supply of eggs.[†]

Parenthetically, we should note that the natural occurrence of embryo and fetal loss and wastage does not necessarily or automatically justify all deliberate, humanly caused destruction of fetal life. For example, the natural loss of embryos in early pregnancy cannot in itself be a warrant for deliberately aborting them or for invasively experimenting on them *in vitro,* any more than stillbirths could be a justification for newborn infanticide. There are many things that happen naturally that we ought not to do deliberately. It is curious how the same people who deny the relevance of nature as a guide for evaluating human interventions into human generation, and who deny that the term "unnatural" carries any ethical weight, will themselves appeal to "nature's way" when it suits their purposes.[††] Still, in this present matter, the closeness to natural procreation—the goal is the same, the embryonic loss is unavoidable and not desired, and the amount of loss is similar—leads me to believe that we do no more intentional or unjustified harm in the one case than in the other, and practice no disrespect.

But must we allow *in vitro* unimplanted embryos to die? Why should they not be either transferred for "adoption" into another infertile woman, or else used for investigative purposes, to seek new knowledge, say about gene action? The first option raises questions about the nature of parenthood and lineage to which I will return. But even on first glance, it would seem likely

[†]There is a good chance that the problem of surplus embryos may be avoidable, for purely technical reasons. Some researchers believe that the uterine receptivity to the transferred embryo might be reduced during the particular menstrual cycle in which the ova are obtained, because of the effects of the hormones given to induce superovulation. They propose that the harvested *eggs* be frozen, and then defrosted one at a time each month for fertilization, culture, and transfer, until pregnancy is achieved. By refusing to fertilize all the eggs at once—i.e., not placing all one's eggs in one uterine cycle—there will not be surplus *embryos,* but at most only surplus eggs. This change in the procedure would make the demise of unimplanted embryos *exactly* analogous to the "natural" embryonic loss in ordinary reproduction.

[††]The literature on intervention in reproduction is both confused and confusing on the crucial matter of the meanings of "nature" or "the natural," and their significance for the ethical issues. It may be as much a mistake to claim that "the natural" has *no* moral force as to suggest that the natural way is best, because natural. Though shallow and slippery thought about nature, and its relation to "good," is a likely source of these confusions, the nature of nature may itself be elusive, making it difficult for even careful thought to capture what is natural.

to raise a large objection from the original couple, who were seeking a child of their own and not the dissemination of their "own" biological children for prenatal adoption.

But what about experimentation on such blastocysts and early embryos? Is that compatible with the respect they deserve? This is the hard question. On balance, I would think not. Invasive and manipulative experiments involving such embryos very likely presume that they are things or mere stuff, and deny the fact of their possible viability. Certain observational and noninvasive experiments might be different. But on the whole, I would think that the respect for human embryos for which I have argued—I repeat, not their so-called right to life—would lead one to oppose most potentially interesting and useful experimentation. This is a dilemma, but one which cannot be ducked or defined away. Either we accept certain great restrictions on the permissible uses of human embryos or we deliberately decide to override—though I hope not deny—the respect due to the embryos.

I am aware that I have pointed toward a seemingly paradoxical conclusion about the treatment of the unimplanted embryos: Leave them alone, and do not create embryos for experimentation only. To let them die "naturally" would be the most respectful course, grounded on a reverence, generically, for their potential humanity, and a respect, individually, for their being the seed and offspring of a particular couple who were themselves seeking only to have a child of their own. An analysis which stressed a "right to life," rather than respect, would of course lead to different conclusions. Only an analysis of the status of the embryo which denied both its so-called "rights" *or* its worthiness of all respect would have no trouble sanctioning its use in investigative research, donation to other couples, commercial transactions, and other activities of these sorts.

***

It is hard to get confident people to face unpleasant prospects. It is hard to get many people to take seriously such "soft" matters as lineage, identity, respect, and self-respect when they are in tension with such "hard" matters as a cure for infertility or new methods of contraception. It is hard to talk about the meaning of sexuality and embodiment in a culture that treats sex increasingly as sport and that has trivialized the significance of gender, marriage, and procreation. It is hard to oppose Federal funding of baby-making in a society which increasingly demands that the Federal government supply all demands, and which—contrary to so much evidence of waste, incompetence, and corruption—continues to believe that only Uncle Sam can do it.

And, finally, it is hard to speak about restraint in a culture that seems to venerate very little above man's own attempt to master all. Here, I am afraid, is the biggest question and the one we perhaps can no longer ask or deal with: the question about the reasonableness of the desire to become masters and possessors of nature, human nature included.

Here we approach the deepest meaning of *in vitro* fertilization. Those who have likened it to artificial insemination are only partly correct. With *in vitro* fertilization, the human embryo emerges for the first time from the natural darkness and privacy of its own mother's womb, where it is hidden away in mystery, into the bright light and utter publicity of the scientist's laboratory, where it will be treated with unswerving rationality, before the clever and shameless eye of the mind and beneath the obedient and equally clever touch of the hand. What does it mean to hold the beginning of human life before your eyes, in your hands—even for 5 days (for the meaning does not depend on duration)? Perhaps the meaning is contained in the following story:

Long ago there was a man of great intellect and great courage. He was a remarkable man, a giant, able to answer questions that no other human being could answer, willing boldly to face any challenge or problem. He was a confident man, a masterful man. He saved his city from disaster and ruled it as a father rules his children, revered by all. But something was wrong in his city. A plague had fallen on generation; infertility afflicted plants, animals, and human beings. The man confidently promised to uncover the cause of the plague and to cure the infertility. Resolutely, dauntlessly, he put his sharp mind to work to solve the problem, to bring the dark things to light. No secrets, no reticences, a full public inquiry. He raged against the representatives of caution, moderation, prudence, and piety, who urged him to curtail his inquiry; he accused them of trying to usurp his rightfully earned power, of trying to replace human and masterful control with submissive reverence. The story ends in tragedy: He solved the problem but, in making visible and public the dark and intimate details of his origins, he ruined his life, and that of his family. In the end, too late, he learns about the price of presumption, of overconfidence, of the overweening desire to master and control one's fate. In symbolic rejection of his desire to look into everything, he punishes his eyes with self-inflicted blindness.

Sophocles seems to suggest that such a man is always in principle—albeit unwittingly—a patricide, a regicide, and a practitioner of incest. We men of modern science may have something to learn from our forebear, Oedipus. It appears that Oedipus, being the kind of man an Oedipus is (the chorus calls him a paradigm of man), had no choice but to learn through suffering. Is it really true that we too have no other choice?

SECTION C

The New Genetics and the American Future

The announcement of the first cloned mammal in 1997 and the mapping of the human genome in 2000 prompted a fresh wave of reflection on how the new genetics is transforming man, society, and politics, and how the modern idea of technological progress may need sober reconsideration.

According to historian Gertrude Himmelfarb, "genetic utopianism" is the latest chapter in the Enlightenment drama of progress and hubris, of great dreams that often end in great nightmares. "Even those of us who do not conceive of immortality," she writes, "aspire to a degree of perfectibility, a control over ourselves and our progeny, and ability to manipulate mind, body, and nature, that no longer seems utopian because it is the practical agenda of scientists and technicians rather than the fancy of philosophers and visionaries." But as with the utopianisms of old, she argues, the result is likely to be more harm than good, a dehumanization of man rather than his rational improvement.

Francis Fukuyama, the political theorist and author of *The End of History and the Last Man*, considers the historic significance of the human genome project and the biotech revolution, which he believes may have the power to "restart" history. He suggests that man's coming biological powers will reshape the longstanding debate over "nature" vs. "nurture." It will force modern societies to make complex, perhaps impossible distinctions between "therapy" and "enhancement." And it could potentially reawaken the political project of creating a "new man," but this time with the actual power to change human nature that earlier utopianisms lacked.

Gilbert Meilaender, the theologian and bioethicist, reflects on the project of "designing our descendants"—a project we have already begun, he says, with selective abortion of "undesirable" fetuses. But why design our descendants at all? How would we design them if we could? According to Meilaender, if we sought to design children with the virtues of prudence, justice, courage, and temperance, and the virtues of faith, hope, and love, such children would have the wisdom not to design their own children. The project of designing our descendents "would bring itself to an end." In contrast, those—perhaps like us, perhaps like all modern societies—who fail to grasp and live by the reality of human limitation, open themselves up to evil: to intolerance bred of false compassion, to lack of acceptance for the "imperfect," to technological despotism in the absence of faith and wisdom.

In the final essay in this section, Adam Wolfson considers the character of our reigning political philosophies—liberalism and conservatism—in a brave new world. He suggests that both liberals and conservatives, for different reasons, have generally embraced the new genetics: liberals because genetic enhancements promise to expand autonomy and equality; conservatives because they promise to meet the market demand for smarter children and better health care. And while there are many in society who reject the new genetics—religious and social conservatives on the Right, Greens on the Left—the dominant sensibilities in America are moral libertarianism and faith in technology. Neither alone, he argues, offers any reliable ground for defending man against those who seek, and may soon be capable of, remaking him.

TWO CHEERS (OR MAYBE JUST ONE) FOR PROGRESS

Gertrude Himmelfarb

If one idea imposes itself upon us as we consider the past thousand years, surely it is the idea of progress. One can hardly begin to take the measure of the extraordinary advances in science, technology, medicine, transportation and communication in this past millennium; or the vast improvements in living and working conditions, health and longevity, education and cultivation, travel and recreation; or the political and social reforms that have brought democracy and liberty to much of the world, making the privileges of the few the rights of the many.

One need not go back 1,000 years to register this spectacular evidence of progress. A century, or half-century, or even a few decades will do. It is not a delusion of contemporaries that we are living in an age of unprecedented change. It is a demonstrable fact, as we have moved from an industrial to a postindustrial society, from a national to a global economy, from a modernist to a postmodernist culture, and, most recently, from a technological to a "hypertechnic" world of seemingly limitless potentialities, where almost anything thinkable becomes possible and almost everything possible probable.

Our present situation brings to mind another period that was intoxicated with the idea of progress. For the Enlightenment, it was not science or technology that was the warrant of progress; it was reason. Reason did not have to await the discoveries of science, because it was innate both in man and in nature, and had only to be liberated from the tyranny of religion, government

Gertrude Himmelfarb is an historian and essayist and professor emeritus at the Graduate School of the City University of New York. Her many books include *The De-Moralization of Society, On Looking into the Abyss*, and *One Nation, Two Cultures*. This essay was originally published in the *Wall Street Journal* on May 5, 1999.

and tradition. This idea of progress was encapsulated in the utopias of William Godwin and the Marquis de Condorcet.

While the American Founders were developing a "science of politics" to ensure the stability of the republic, Godwin was developing a "science of morals" to ensure the perfectibility of man. The first principle of Godwin's science was reason. Human beings who were perfectly rational would be perfectly moral, hence free and equal. Thus there would be no need for all the oppressive institutions that make up society: laws, contracts, property, religion, marriage, family, schools, armies, prisons.

Almost as a byproduct, perfectibility would bring with it the "total extirpation of the infirmities of our nature"—disease, sleep, languor, anguish, melancholy, resentment, even death itself. Godwin did not commit himself to the idea of immortality as a certainty, only as a probability; what was certain was the infinite prolongation of life. This happy state of affairs would not lead, as one might fear, to an onerous overpopulation, because men, being perfectly rational, would be liberated from emotion and passion. Thus there would be no sexual desire and no propagation. "The world," Godwin was pleased to report, "will be a people of men, and not of children. Generation will not succeed generation, nor truth have in a certain degree to recommence her career at the end of every thirty years."

This was surely the utopia to end all utopias—until our present utopia. Today, by means of genetic and embryonic engineering, organ replacement, and, now, cloning, we can look forward to a time when we will be able to create individuals to specification and, in principle at least, sustain them indefinitely. "This is the first time we can conceive of human immortality," says the head of a thriving biotech company, who is a respected scientist but clearly no historian. Even those of us who do not conceive of immortality aspire to a degree of perfectibility, a control over ourselves and our progeny, an ability to manipulate mind, body and nature, that no longer seems utopian because it is the practical agenda of scientists and technicians rather than the fancy of philosophers and visionaries.

While "ethicists" debate the implications of all this, the rest of us reread Aldous Huxley's "Brave New World" and marvel at his prescience. We may take comfort in the fact that our brave new world, unlike his, is not totalitarian; indeed, it is eminently free and uninhibited, catering to the wills and desires of individuals. It does, however, share with his the overriding idea of progress. "Progress is lovely" is the mantra endlessly repeated for the edification and indoctrination of the inhabitants of Huxley's world. In the heady atmosphere we live in today, it threatens to become our mantra as well.

Yet progress is not always lovely. Sometimes it is notably unlovely. The perfect can be the enemy of the good, as the old adage has it. Immoderate desires and aspirations make us impatient with the frailties of human nature and the contingencies of social institutions, prompting us to try to overcome these imperfections by creating a new man, a new society and a new polity. Such enterprises, history reminds us, are fraught with peril. We recall that Godwin's work, greatly admired by the English enthusiasts of the French Revolution, appeared precisely at the moment when Robespierre's "Reign of Virtue" mutated into a Reign of Terror. And Condorcet's "Sketch of the Progress of the Human Spirit," celebrating the rationality of man, infinite progress of mankind, and unlimited extension of life, was written while he was hiding from the revolutionary police (he supported the revolution but opposed the execution of the king) and was published only after he had died in prison.

This is the great advantage of religious utopias over secular ones. Religious utopias are otherworldly; they preserve the transcendent vision of perfection without seeking to actualize that vision on earth. Secular ones seek to create a utopia on earth, an act of hubris that is almost always fatal, if not to the perpetrators of the utopia, then to its innocent victims.

Contra naturam, the defiance of nature, used to be a sufficient argument for those who were not persuaded by *contra deum*, provoking the wrath of God. But what does it mean today, when we have defied, even violated, nature in so many ways, for good as well as bad? If cloning is against nature, is not also artificial insemination, or in vitro fertilization, or, for that matter, the pill? If we approve of embryonic and genetic research to remedy birth defects, why not cloning to create the perfect child, or the perfect replica of oneself, or of another loved one, or of a much-admired celebrity?

Objections to biotechnology are commonly couched in pragmatic, empirical terms. Cloning, it is argued, may lead to incest and thus to the deformations associated with inbreeding. Or embryonic research may encourage abortions to provide the embryos for experimentation, or even encourage pregnancies intended to be aborted. These, and a host of similar objections, are serious enough to give us pause.

But the ultimate question is how far we may go in defying nature without undermining our humanity. What does it mean to create human beings *de novo* and to specification? What does it mean for human beings, who are defined by their mortality, to entertain, even fleetingly, even as a remote possibility, the idea of immortality? It is at this point that contra naturam rises to the level of contra deum.

To raise these questions is in no way to reject science and technology or to belittle their achievements. It is not contra naturam to invent labor-saving devices and amenities that improve the quality of life for masses of people, or medicines that conquer disease, or contrivances that allow disabled people to live, work and function normally. These enhance humanity; they do not presume to transcend it.

But science and technology, like progress itself, can be morally equivocal. It was not a millennium ago but in this very century that we experienced one of the most monstrous events in human history, the Holocaust, and discovered, not for the first time, that both science and technology can be put to the most heinous uses. We have also been obliged to reconsider the Enlightenment, which bequeathed to us many splendid achievements but also some dangerous illusions.

In our post-Enlightenment world, we have had to relearn what ancient philosophy and religion had taught us and what recent history has brought home to us: that material progress can have an inverse relationship to moral progress, that the most benign social policies can have unintended and unfortunate effects, that national passions can be exacerbated in an ostensibly global world and religious passions in a supposedly secular one, and that our most cherished principles (liberty, equality, fraternity, even peace) can be perverted and degraded—that, in short, progress in all spheres, not only in science and technology, is unpredictable and undependable.

This may be the lesson of the millennium. Progress, yes, but a modest, cautious, amelioratory progress, chastened by the experiences of history and guided by a sense of human limits as well as possibilities.

A MILESTONE IN THE CONQUEST OF NATURE

Francis Fukuyama

Yesterday's joint announcement that Celera and the Human Genome Project had completed a rough sequencing of the human genome can be both overestimated and underestimated in its significance.

The commercial hype surrounding the announcement is enormous, with some going so far as to suggest that this advance will have immediate consequences for health and happiness. This is a great overstatement. What the scientists have done is to transcribe, in computer-searchable form, an extremely long book written in a foreign language, without providing a dictionary. We currently know only a very few words and phrases of the language, as well as the fact that as much as 95 percent of the text is gibberish.

An immense task of translation still lies ahead. Other biotech companies and researchers will have to identify genes within the genome, explain what proteins they produce, and then identify how those proteins affect things we care about—like a propensity for breast cancer, intelligence, Alzheimer's or longevity. The private companies trying to "patent" genes aren't patenting the genome itself, but rather their translations and interpretations of it. The joint announcement, therefore, marks only the beginning of an enormously long and difficult research effort.

Still, no one should underestimate the significance of the announcement. Many people believe that decoding the human genome will primarily lead to

Francis Fukuyama is a political theorist and Bernard Schwartz Professor of International Political Economy at the Paul H. Nitze School of Advanced International Studies, Johns Hopkins University. His many books include *The End of History and the Last Man*, *The Great Disruption*, and a forthcoming work on the political implications of biotechnology. This essay was originally published in the *Wall Street Journal* on June 27, 2000.

better ways of developing drugs. They see its main problems as issues like medical privacy or the insurability of people born with certain genetic markers. Important as these concerns are, they pale in comparison to issues we will face in the future. Unpacking the human genome is a significant milestone in the ongoing, 500-year modern project of (in Francis Bacon's words) seeking "the relief of Man's estate" through the progressive conquest of nature.

Through most of human history, the nature that we have been seeking to master has been that of our external environment: floods, plagues, droughts, scarcities. But the greatest constraints on human freedom are those imposed by our own human natures. We are mortal, selfish, irrational and overly emotional, limited in intelligence and perception, prone to violence and aggression, and blindly loyal to family and friends.

The decoding of the genome will help settle many of the "nature vs. nurture" questions that have dogged philosophy since the ancient Greeks, and that are at the core of any number of today's public-policy debates. Are men and women truly different psychologically, or is it just a matter of social conditioning? Is homosexuality a congenital or an acquired condition? To what extent is intelligence inherited, or is it something that can be improved through a better environment? Are there significant differences between racial and ethnic groups beyond skin and hair color? Each of these positions has fierce partisans, but without being able to link specific genes to specific conditions or behaviors, the arguments remain largely speculative.

The answer to these questions is never a simple "either-or," but rather a statement about the effect of certain factors on the "variance" in human behavior. In the middle of the 20th century, social scientists believed that culture and environment counted for nearly 100 percent of the variance, and biology almost none. With the rise of disciplines like behavioral genetics (largely based on studies of twins), the tide has been shifting steadily in favor of genetic factors. The ability to actually link behavior to genes on a molecular level will accelerate this trend and make the findings more precise. We probably won't like the answers, because we may find that we are much less free to choose our destinies as we would like to believe.

Given that what Marx labeled the "realm of Nature" looms large as a constraint on human aspirations, it seems almost inevitable that we will eventually seek to use genetic knowledge to actively reshape human nature. This could take many forms, from wealthy parents creating "designer children" with superior looks and intelligence, to an egalitarian state trying to remedy natural inequality through a new form of eugenics. Once we better understand the genetic sources of behavior, we will be able to develop powerful new tools to better control it. The way is then open to superceding the human race with something different.

In the face of statements like these, corporate public-relations types begin to get nervous that their own rhetoric is being taken too seriously. They remind people that biotechnology is not about re-engineering people, but about curing disease and helping people to live happier and healthier lives. It is to be used for therapeutic purposes, not to violate nature but to help people to live more in accordance with it. Human behavior is in any event very complex, they say, and the likelihood that we will be able to modify it significantly is very low.

But as the bioethicist Leon Kass has pointed out, the line between therapy and enhancement won't hold. There is no strict dividing line between sickness and health when it comes to conditions we label "pathological." A condition diagnosed as "attention deficit-hyperactivity disorder" by one physician might seem like normal boyish high spirits to another. Suppose, says Mr. Kass, that biotechnology gives us the capability to change height, and thereby cure dwarfism. Who will tell the parent with a child in the fifth percentile for height that they shouldn't be allowed to make their child taller, given the clear advantages height confers? And if there is no objection to giving the therapy to someone in the 5th percentile, why not give it to someone in the 50th?

One might be tempted to ask why we shouldn't use biotechnology for enhancement. Parents want the best for their children, whether it is a matter of height, intelligence, looks or social adjustment. Who is going to tell them this is wrong? Biotechnology, we might say, will help the human race get better.

It may be there is nothing wrong with the hope that biotechnology will deliver to us a future that is full of more promise than any utopia we've heretofore dreamed up. But there is a nagging ground for worry about the wisdom of this path, especially given the potential consequences for politics.

The institutions of our current liberal democratic order, from the family to the market to democracy itself, are based on the fact that human nature is one way and not another. All of the radical revolutionary movements of the past three centuries, from the French Revolution to the Bolshevik, Chinese, and Cambodian upheavals, in contrast, were premised on the belief that human nature was highly plastic and shapeable through social policy. If people didn't conform to the revolutionaries' prior assumptions, they could be made to do so through labor camps, agitprop and re-education.

The belief that human behavior could be shaped by social engineering had horrific consequences, and the spread of liberal democracy through large parts of the world at the expense of socialism at the end of the twentieth century in large measure reflects the recognition that it couldn't be made to work. Socialism foundered, in a sense, on the shoals of a human nature that wouldn't let utopian planners do as they wished.

So this is the question posed by the achievement of Celera and the Human Genome Project: What kind of politics will a presumed future knowledge of the genome make possible? Could it be that the technologies for social engineering pioneered during the twentieth century didn't work only because they were too crude, but that in the future we will have biotechnology to do the job better? Eugenics has been back on the table ever since amniocentesis made possible the abortion of children with severe birth defects. Charles Murray recently suggested that in the future, eugenics will be championed by the left rather than the right, in an effort to remedy natural inequality. What kinds of passions will be released when the stakes become a society's genetic future?

Amid all the celebrating on Wall Street and in the scientific community after the joint announcement, this is something to think about.

DESIGNING OUR DESCENDANTS

Gilbert Meilaender

In one of the classic early discussions about the possible uses of advancing genetic knowledge to control and reshape human life, Paul Ramsey, more than thirty years ago, wrote the following:

> I . . . raise the question whether a scientist has not an entirely "frivolous conscience" who, faced with the awesome technical possibility that soon human life may be created in the laboratory and then be either terminated or preserved in existence as an experiment, or, who gets up at scientific meetings and gathers to himself newspaper headlines by urging his colleagues to prepare for that scientific accomplishment by giving attention to the "ethical" questions it raises—if he is not at the same time, and in advance, prepared to stop the whole procedure should the "ethical finding" concerning this fact-situation turn out to be, for any serious conscience, murder. It would perhaps be better not to raise the ethical issues than not to raise them in earnest.

My aim in this brief essay is to invite thought about *present* uses of genetic screening and the attitude we ought *now* to have toward the project of shaping our children, lest a focus on *future* possibilities should tempt us here and now to "an entirely 'frivolous conscience.'"

No one doubts that genetic advance will, in good time, enable us to find therapies for at least some of the devastating diseases whose causes are, in whole or part, genetic. Research is taking place on many different fronts.

Gilbert Meilaender is a theologian and bioethicist and holds the Richard and Phyllis Duesenberg Chair in Christian Ethics at Valparaiso University. His many publications include *The Theory and Practice of Virtue, Body, Soul, and Bioethics,* and *Things That Count.* This essay was originally published in *First Things* in January 2001.

Xeno-transplantation—that is, transplantation of animal organs, probably pig organs, into human beings—is the focus of much work right now. Genetic therapy—in which a functioning copy of a gene is added, or, even more desirable were it possible, a defective gene is removed and replaced with a functioning version—is the holy grail of research. Although progress on the therapeutic front has been slower than many had predicted, it remains the focus of much research and many hopes. Perhaps more dramatically still, what is called germ-line therapy—an alteration not of the somatic cells of the body but of the germ cells, the sex cells by which traits are passed on to future generations—is increasingly defended and may become possible. Just as mind-boggling is the work being done to culture embryonic stem cells in order to grow organs or tissues for transplant. Because such stem cells are essentially immortal—they simply regenerate themselves—this research may even hold out the hope of retarding the aging process, which some people, at least, think desirable.

In short, from countless different angles the pace of research invites us to reflect upon the project of controlling and reshaping human beings or, more broadly still, human nature—all in the name of relieving suffering. Our language regularly invites us to view this project in favorable terms. Consider, for example, the verbal formulations I might have used here. I might have said that researchers are "progressing" or "advancing" in the treatment of genetic disease and applications of molecular biology in medicine. It would have been surprising and counterintuitive had I written that the sorts of possibilities mentioned above indicate that researchers are "regressing" or "retreating." We can scarcely imagine that increased ability to relieve suffering or eliminate defect and disease could be anything other than progress and advance.

At the same time, everyone also acknowledges that new techniques could be misused, even though few agree on exactly what would constitute such misuse. So, for example, we have distinguished between somatic cell and germ cell therapy, approving the former and disapproving the latter. But anyone reading the literature will surely note how that line has begun to break down of late, as an increasing number of bioethicists seem willing to defend germ-line interventions. Or, we have drawn a line between therapy and enhancement—embracing the former while disapproving the latter. But, again, anyone reading the literature will note how this line too has begun to blur. In response to that blurring we may try to distinguish between health-related and non-health-related enhancements—between, for example, enhancing the body's ability to fight certain cancers and enhancing memory or, even, kindness—but it would surely be naive of us to suppose that the pressure to blur this line will not also be enormous.

To the degree that we as a people have lost the capacity to draw lines, to decide what should and should not be done, we are forced to take refuge in virtue ethics. That is, if we cannot definitively say which acts should or should not be done, perhaps we can trust people of good character to use these new techniques without abusing them. Yet, of course, *we* are the people who will be using the advances in genetics and whose wisdom and virtue must be trusted. What kind of people are we? To answer that question we need to think not about what may be possible in the future but, rather, about what is done in the present. We need to focus not on future subjunctives but on present indicatives.

And our present condition is this: we have entered a new era of eugenics. That science which attempts to improve the inherited characteristics of the species and which had gone so suddenly out of fashion after World War II and the Nazi doctors now climbs steadily back toward respectability. Eugenics becomes respectable again insofar as it promises to relieve suffering, as it claims for itself the virtue of compassion. The new eugenics has, however, a distinctly postmodern ring. In the heyday of eugenics early in the twentieth century, its proponents had in mind government-sponsored programs that might even involve centralized coercion. Thus, for example, as Bryan Appleyard, a columnist for the *Sunday Times* of London, has noted, "A Geneticist's Manifesto," signed in 1939 by twenty-two very eminent American and British scientists, "called for the replacement of the 'superstitious attitude towards sex and reproduction now prevalent' with 'a scientific and social attitude' that would make it 'an honor and a privilege, if not a duty, for a mother, married or unmarried, or for a couple, to have the best children possible, both in respect of their upbringing and their genetic endowment.'"

By contrast, the new eugenics comes embedded in the language of privacy and choice. Its two cardinal virtues—almost the only virtues our culture now knows—are compassion and consent. Compassion moves us to relieve suffering whenever possible; consent requires that our compassion be "privatized." A world in which prenatal screening followed by abortion of children diagnosed with defects has become a routine part of medical care for pregnant women—that is to say, our world—is one into which eugenics enters not through government programs but precisely as government removes itself from what is seen as entirely a private choice.

I am not invoking the "Nazi analogy," which has been disputed as often as it has been invoked in bioethical argument. I am not claiming that we find ourselves on a slippery slope, at the foot of which might lie Nazi-like deeds. Instead, I simply note how easily we may deceive ourselves about what we do here and now, how subjectively well-meaning people may approve objective evil. Just that, in fact, is the most terrifying point of Robert Jay Lifton's great work, *The Nazi Doctors*. He invites us to see within ourselves—good people whom we suppose ourselves to be—the possibility of great evil. "We thus find ourselves returning," Lifton writes near the end of his discussion of the "doubling" that made it psychologically possible to be a Nazi doctor, "to the recognition that most of what Nazi doctors did would be within the potential capability—at least under certain conditions—of most doctors and of most people." To read Lifton's account is, for the most part, to read of good and ordinary people in the grip of an ideology who suppose themselves to be engaged in a compassionate and healing endeavor. Something like that, Lifton suggests, is an almost universal possibility. And something like that, I am suggesting—something very different because cloaked in the language of privacy, yet not so different because it sanitizes and medicalizes as healing the killing of "defectives"—is by far the most common present use of genetic "advance."

This is our present situation. The day may come when we can treat and cure prenatally or postnatally many genetic diseases; however, for the moment we can diagnose prenatally far more than we can treat. In the meantime, therefore, we screen and abort. For now that is essentially the only "treatment" for illness diagnosed prenatally. We know more and more about the child *in utero*; hence, people quite naturally seek and use such knowledge in order to select the babies they desire and abort those they do not want. This is the new eugenics—which relies not on government coercion but on private choice and desire, on the commodification of children. Thus, Appleyard writes, "We could not now respectably speak of 'the improvement of the race' or of 'selective breeding'—the terminology of the old eugenics—but we do speak of the 'quality of life' and assess our children in consumerist terms." Not only is the meaning of childhood distorted but the meaning of parenthood as well. Selective abortion means selective acceptance. The unconditional character of maternal and paternal love is replaced by choice, quality control, and an only conditional acceptance.

People who have chosen or have been taught to think this way are the people who will be deciding what constitutes proper use or misuse of advancing genetic technology. We run the risk of cultivating frivolous consciences, therefore, if we pay attention to the moral conundrums of future subjunctives while studiously ignoring present indicatives. For as long as we are willing to rely on the routinized practice of prenatal screening followed by selective abortion we are not people who should be trusted to design their descendants.

Suppose, however, that we could be so trusted and that the capacity to design and shape our children, to shape their nature and character, were really ours. What sort of people should we aim to create? Taking my inspiration from a short piece written by Alasdair MacIntyre more than two decades ago ("Seven Traits for the Future," *Hastings Center Report*, February 1979), I suggest that we ponder this question briefly. An obvious way to answer the question is to think in terms that our moral tradition has taught us, in terms of the four cardinal and the three theological virtues. We should aim to design children who would be characterized by prudence, justice, courage, and temperance—and, in addition, by faith, hope, and love.

Within the Western moral tradition, so nicely articulated by Josef Pieper in *The Four Cardinal Virtues*, the virtue of prudence means something quite different from our use today of a word such as "prudential." For us, in fact, there is a certain tension between being good and being prudent. In the longer view of the tradition, however, the virtue of prudence enables us to see things as they really are. Not just as we would like them to be, or fear that they must be, or greedily hope they may be—but as they truly are.

Two things follow. On the one hand, this implies that nature itself—apart from our own shaping activity—has order and form to be discerned by the prudent man or woman. Prudence seeks to conform to the reality of things and does not suppose that our humanity consists only in our power of mastery over nature. Hence, the first task, as MacIntyre suggested, is not to change the world but to understand and interpret it.

On the other hand, human prudence can never exhaust or comprehensively understand the meaning and order of our world. MacIntyre illustrates how the limits to our understanding are evident in the very possibility of radical conceptual innovation. Relying on Karl Popper, he asks what it would have been like had one of our ancestors predicted the invention of the wheel. In order to explain his prediction this ancestor would have had to describe axle, rim, and spokes—in short not just to predict the invention of, but, in fact, to invent the wheel. The lesson being: we simply cannot predict with any precision what the future may be like, what innovations may appear; and we will want children who can accept the limits of our knowledge, who know that not everything is within our control.

To see the world rightly is, among other things, to see the difference between "mine" and "thine"—that is, to be just. Justice is not yet love; it is life *with*, not life *for*, the neighbor. Without denying our fellow humanity, justice maintains distance between us, lest the life of one should be entirely absorbed within the aims and purposes of another. Hence, to be just is to respect the rights of others, to recognize the claim upon us of their equal dignity. The denial of that common humanity is, as Pieper writes, "the formal justification for every exercise of totalitarian power." Such power, of course, can be exercised not only synchronically but also diachronically—across generations.

Justice requires, at least in some times and places, courage. For there will be moments when it is injurious to one's own interests or, simply, dangerous to be just. Courage is a necessary human virtue precisely because we are creatures always vulnerable to injury—the greatest of which, of course, is death. We have tended to picture courage as a martial virtue needed by those who must face enemies and fight. But Pieper reminds us that in our world, and perhaps in any world, it is needed as much by those who only suffer—not those who attack, but those who endure. It is needed if we are, one day, to accept the appropriateness of our own death, to acknowledge that a just affirmation of the claims of others means that we must recognize that our own time and place is not the master of every time and place, that we must give place to those who come after us.

To be that courageous in the service of justice cannot be possible for one whose inner self is not fundamentally in right order and at peace. Lacking such order, we are bound to grasp for what we desire at any moment, to flinch when sacrifice of our own desires seems needed for the sake of others. We should, therefore, want our children to be not only prudent, just, and courageous, but also temperate.

High as such an ideal of character is, however, Christians will not rest content with an ideal that can and has been known apart from Christ. They will also want the threefold graces of faith, hope, and love to be formed in their children. Faith is "the conviction of things not seen." Because we cannot fully see the way ahead, because, as I noted above, we must always live with uncertainty about the future, we are unable to secure our own lives. We exercise control of various sorts, we improve the human condition, but mastery eludes us. Hence, we must want our children to know the limits of their power, to seek wisdom more than power. Indeed, if our faith seeks its security finally in God, our life must be oriented toward One who most emphatically is not within our control. As philosopher Robert Meagher writes, describing St. Augustine's view: "It seems that one may either strive to want the right thing, or strive to have what one wants. The search for wisdom somehow involves the renunciation of power, the renunciation of possession, while the search for power somehow involves the renunciation of wisdom, since it presupposes the appropriateness of what it is striving to attain."

Because faith requires that we live without trying to secure our own future, it needs to be joined with the virtue of hope. We must hope and expect that God can complete what is incomplete in our own strivings, especially when, in order to live justly, we refrain from achieving good that can be gotten only by evil means. "You might have helped me," future generations may say, "had you only been willing to dirty your hands a bit." There is no reply to such a charge without an appeal to hope, without a sense of the limits of our responsibility to do the good. "Above all," Kierkegaard writes, "the one who in truth wills the Good must not be 'busy.' In quiet patience he must leave it to the Good itself what reward he shall have, and what he shall accomplish." To be hopeful is to adopt the posture of one who waits, who knows his fundamental neediness and dependence; for, after all, were the good for which we wait at our own disposal, there would be no need for hope.

There remains love, the greatest of the virtues. Love signifies approval, and, Josef Pieper writes, "the most extreme form of affirmation that can possibly be conceived of is *creatio*, making to be." Hence, our own love mirrors the creative love of God, which bestows on us the divine word of approval. Love therefore is, in Pieper's words, a way of turning to another and saying, "It's good that you exist; it's good that you are in this world." As we hope to become people who can love our own children in this way, so we would want them, in turn, to be people who can love as they have been loved—with an affirmation that is not conditioned upon the qualities of the loved one.

We can say, by way of summary, that, were we to undertake the project of designing our descendants, we should want them to be people who do not think the natural world infinitely malleable to their projects; who reckon from the outset with limits to their own knowledge of and control over the future; who respect the equal dignity of their fellows and do not seek to coopt others as means to their own (even if good) ends; who acknowledge even their own death, the ultimate of limits; who are prepared to subordinate their needs to the good of others; who are more disposed to seek wisdom than power; who know that the good is not finally at their own disposal; and who live in a manner that says to others, "It's good that you exist."

My argument has come at the question of designing our descendants in two stages. I first suggested that people who live as we do—who have accepted as useful the routinized practice of prenatal screening—are people who have no business attempting to reach out and shape future generations.

But, second, I asked what kind of descendants we should seek to create if we were people fit to undertake such a task. If we are drawn to the description I have given in terms of the cardinal and the theological virtues, we should be able to see, from a quite different angle, why designing descendants is a project we ought not undertake. I said earlier that I took the inspiration for this idea from Alasdair MacIntyre, and I can do no better than repeat here the words with which he concluded his own exploration of the traits we should want our children to have.

> If in designing our descendants we succeeded in designing people who possessed just those traits that I have described, . . . what we would have done is to design descendants whose virtues would be such that they would be quite unwilling in turn to design *their* descendants. We should in fact have brought our own project of designing descendants to an end.
>
> It turns out then that my argument has immediate practical consequences. For if we conclude that the project of designing our descendants would, if successful, result in descendants who would reject that project, then it would clearly be better never to embark on our project at all. Otherwise we shall risk producing descendants who will be deeply ungrateful and aghast at the people—ourselves—who brought them into existence.

POLITICS IN A BRAVE NEW WORLD

Adam Wolfson

By all accounts, we are on the cusp of a great technological revolution or revolutions. In the last several months, *Time*, *Newsweek*, *Wired*, the *New York Times Magazine*, even the stuffy, high-brow *Partisan Review* have run cover stories on the technological marvels about to transform our lives. None of this apparently is science fiction but what leading scientists in their respective fields are predicting as fact. Some of the changes will be of the James Bond variety, like auto-piloted cars and nonlethal phaser guns. Other inventions will have broader effects: New virtual-reality games and the Internet will complete a process that TV inaugurated, the metamorphosis of our civilization into one increasingly driven by images and sentiment rather than words and thought. And other changes in the offing will be even more dramatic, going to the very heart of what it means to be a human being. Discoveries just over the immediate horizon in human genetics and computers threaten (or promise, depending upon your perspective) to usher in, as strange as it may sound, a "posthuman" era.

As or even more interesting than the foretold scientific breakthroughs, however, is our collective reaction to them. Generally, the American public is quiescent, when it is not celebratory, and many leading liberal and conservative intellectuals seem relatively untroubled. Gone apparently are the days when conservatives and liberals both worried about the dangers of the Machine. A half century ago, Richard Weaver, a revered and iconic figure on the

Adam Wolfson is executive editor of *The Public Interest*. His many essays and articles have appeared in the *American Scholar*, *Commentary*, *National Review*, the *Review of Politics*, *Policy Review*, the *Wall Street Journal*, the *Washington Post*, the *Weekly Standard*, the *Wilson Quarterly*, and others. This essay was originally published in *The Public Interest,* Winter 2001.

American Right, famously denounced "the gods of mass and speed" in his conservative classic *Ideas Have Consequences*. And one need only mention such a stalwart of the Left as Herbert Marcuse, who in his sixties manifesto, *One-Dimensional Man*, attacked "technological rationality" as "the great vehicle of better domination, creating a truly totalitarian universe," to remember that liberals once cast a cold eye toward technology. As literary critic Leo Marx has shown, censure of technology has had a long and venerable history in America, stretching from Thomas Jefferson and Nathaniel Hawthorne to F. Scott Fitzgerald and the 1960s counterculture.

"Somewhere along the way, however [as the editors of the *New York Times Magazine* recently wrote approvingly], all that doom and gloom drifted out to sea, pushed away not by blue-sky idealism but by a temperate, practical faith in the idea that machines are nothing to be afraid of, that we can use them when, where and how we want to use them and that technology might actually improve our lives." Indeed, there can be no more definitive measure of our embrace of technology than our reaction to advances in the field of genetic engineering, where the possibilities are truly consequential. Lee M. Silver, a molecular biologist and neuroscientist at Princeton University, predicts that by mid century genetic engineering will have become sufficiently feasible, safe, and efficient that we will possess "the power to change the nature of humankind." And the founder of sociobiology, Harvard professor Edward O. Wilson, estimates that within a few decades, we will enter an era of "volitional evolution," having gained the ability to "alter not just the anatomy and intelligence of the species but also the emotions and creative drive that compose the very core of human nature." If anything should occasion introspection and pause, if not doom and gloom, surely this is it. Yet even when the stakes could not be higher, when we are about to embark on the (re)creation of man, liberalism and conservatism seem to have put out the welcome mat. It is worth asking why.

LIBERALS

As it happens, there is a critical liberal response to technology's encroachment on human life, and it has been convincingly made by William A. Galston, a former advisor in the Clinton administration and the author of *Liberal Purposes*. In Galston's view, we intuitively sense that modern technology can, on the one hand, supplement or complete human nature, and on the other hand, violate it. Reading machines for the blind would be an example of the former, a technology that allows the disabled to lead more fully human lives; while medical technologies that prolong dying and suffering detract from human dignity by riding roughshod over natural limits. Galston himself contrasts a proper, liberal use of genetic treatment—which would, for example, seek "to nip Down's syndrome in the bud, that is, to eliminate natural defects in a manner that safeguards the life of the fetus"—with an illiberal form of genetic engineering that would favor certain traits over others or even introduce new traits into the species. The question is whether liberalism has within itself the resources and depths to ground such distinctions. Galston contends that, at least in theory, a neo-Kantian understanding of human freedom does—that it can "help draw the line between technology supportive of our dignity and technology that erodes dignity by treating us as means alone."

Galston perhaps shows that liberalism, properly understood, can help guide us through the technological thicket. But as he acknowledges, political philosophies in the real world are rarely properly understood, much less properly practiced. I would point out that the collapse of liberalism's standard against obscenity, culminating in the Supreme Court's declaration in 1971 that "one man's vulgarity is another man's lyric," does not inspire optimism. Liberals who cannot tell the difference between art and obscenity are not likely to rise to the more difficult challenge of differentiating technologies that fulfill our nature from those that destroy it.

And in fact the liberal theorist Ronald Dworkin argues not simply that liberalism is neutral on the question of genetic engineering but that it welcomes it. The author of numerous books in legal and political theory, and an important influence on the Supreme Court, Dworkin is, without exaggeration, the leading voice of elite liberal opinion in America. In his latest book, *Sovereign Virtue*, Dworkin derives a eugenic imperative out of liberalism's principles of equality and autonomy. Regarding the equality principle, Dworkin argues "that it is objectively important that any human life, once begun, succeed rather than fail—that the potential of that life be realized rather than wasted—and that this is equally objectively important in the case of each human life." And regarding liberalism's principle of autonomy, Dworkin holds that a person "has a right to make the fundamental decisions that define, for him, what a successful life would be." Taken together, these two liberal principles mandate, in Dworkin's view, a new eugenics.

The first principle—that a life once begun be successful—requires, according to Dworkin, a regime of genetic testing, perhaps even mandatory testing. For if it becomes possible to test for and correct genetic defects in embryo, then our concern for equality, our concern that each life be successful, necessitates that we do all we can to prevent genetically linked diseases. The liberal principles of equality and autonomy also justify, in Dworkin's hands, the more radical program of genetically reengineering the species—of deciding, in Dworkin's blunt words, "which kind of people, produced in which way, there are to be." Here's his reason why: The equality principle "commands" that we make every life a successful one, and thus genetically enhanced IQs and the like, once feasible, are morally required. The autonomy principle meanwhile "forbids" "hobbling the scientists and doctors who volunteer to lead" the new eugenics, for we are not permitted to interfere with the scientist's or any individual's autonomy. If the scientist believes it his mission to transform the human species, we must respect his autonomy, and allow him to proceed.

It might at first seem shocking that a prominent liberal would take this position. What's being defended is a eugenic program to weed out "unsuccessful" human life—a program to be led by an elite corps of scientists accountable to no one but themselves. Yet on reflection we should perhaps not be surprised that liberalism can be interpreted (or misinterpreted) in this way: The Left's belief in Progress leads it to seek the "improvement" of the human race, whether by education or therapy or psychotropic drugs or . . . eugenics. Its devotion to absolute human equality tempts it to achieve this elusive goal if not by social engineering then by genetic engineering. Its espousal of radical autonomy leaves no standard by which the choices of scientists can be questioned, much less circumscribed. And more broadly, its concept of state neutrality regarding human ends (i.e., "lifestyle" choices) prepares the way for a state that is also neutral regarding the (technological) means chosen to reach those ends. Finally, modern liberalism's inveterate hostility toward religion (except Senator Joe Lieberman, of course) goads it, in Dworkin's words, to "play God," "because the alternative is cowardice in the face of the unknown."

CONSERVATIVES

The case of conservative opinion and the new technologies, eugenics in particular, is not so clear-cut. Religious conservatives, of course, have strongly resisted fetal-tissue research as well as other biotech developments. Meanwhile, social conservatives have put forward powerful and convincing critiques of the new eugenics, and of these none stands out more than Leon

Kass's. In Kass's neo-Aristotelian view, the problem with modern science, notwithstanding its success in explaining how things work, is its failure to say anything about human ends or the good for man. Thus Kass has sought, in his own words, "a more natural and richer biology and anthropology, one that does justice to our lived experience of ourselves as psychophysical unities—enlivened, purposive, and open to and in converse with the larger world." His argument is that only by understanding human ends as they are revealed in our "given nature" will we come to see the violation of genetic engineering. Kass has leveled one of the most trenchant philosophic challenges in recent years to the new technologies. Yet it is probably fair to say that he is not representative of modern conservative opinion, and so one must look elsewhere for the *popular* conservative take on technology.

Modern conservatism is characterized by its denunciation of Big Government and its cry of "leave us alone." It is this libertarian view that guides most conservative politicians and policy makers as well as grass-root conservatives. Behind these popular slogans lies a sophisticated theory of the nature of society and the limits of human understanding, a theory developed by F.A. Hayek, who has had an enormous influence on American conservatism. On balance, Hayek's philosophy probably contributes to conservatives' acceptance of the new eugenics, but not without raising important questions along the way.

According to Hayek, society is a "grown" or "spontaneous order," as opposed to a "made" order. By this, Hayek meant that society comes into being and functions without conscious design, and that neither society's origins nor its operations can be comprehended by the human mind. The effort to do so, to construct the ideal social order or the planned economy—an effort that led in the last century to untold suffering and catastrophe—Hayek attributed to the philosophies of Descartes, Bacon, Comte, and Marx. He gave a special name to this new mode—"constructive rationalism"—but he was not the only conservative thinker to be critical of its effects. From Burke to Michael Oakeshott to Milton Friedman runs a critique of the modern rationalist conceit of building utopia.

So where might the Hayekian understanding of social formation and the limits of human understanding lead on the question of the new eugenics? Well, consider for a moment the buzzwords commonly associated with the new eugenics: We are told that our scientists have discovered human nature's "blue print" or its "operating manual"; that our molecular biologists will be able to replace slow, blind evolution with rational control; that they will be able to "custom design" human beings; that they will become the creators of a new race of men; that they will indeed play God. This is the "constructive rationalist" enterprise par excellence! And not incidentally, it received its first impetus from those godfathers of modern science Bacon and Descartes, precisely the philosophers Hayek singled out for opprobrium. Yet if one were to suppose that all this talk of human design, rational planning, scientific expertise, and progress and perfection has led contemporary conservatives to reject the new eugenics one would be wrong. The libertarian magazine *Reason* and the more traditionally conservative *National Review* have both run articles endorsing the new eugenics. And while Charles Murray, the author of *The Bell Curve* and a primer on libertarianism, has strongly criticized a left-wing brand of eugenics, whereby the government implements eugenic programs to improve the IQ of poor children, he has endorsed voluntary genetic manipulations aimed at improving overall physical and mental abilities.

The conservative embrace of the rationalist project of Bacon might be explained by the fact that Hayek and other libertarian thinkers were concerned with the evils of *social* planning, not *biological* planning. It is perhaps the rationalist approach applied to society and politics only that is condemned by them. However, it's not clear that the two are easily separated, for in Hayek's view, the human organism, like society, is a "spontaneous order" (though of a more concrete kind). I would raise then the possibility that some of the strictures Hayek leveled against social planning could equally apply to the case of genetic engineering, though the last word on this must go to Hayek scholars. On a more practical and telling level, the two cannot be separated since the application of "constructive rationalism" in one sphere, genetics, will have, as all acknowledge, far-reaching political and social consequences. What these will be is anyone's guess, but one way or another, rational control and manipulation of the genetic blue print will bring about the triumph of Rationalism in politics.

LIBERTARIANS ALL

This is not to say, however, that modern conservative ideology stands against genetic engineering. Clearly it does not. One reason is its guiding principle of individualism—that individuals are best situated to make choices that concern their well-being. Just as the concept of the "spontaneous order" rules out social planning, so too it rejects the possibility of transcendent moral standards. Values and standards are said to be subjective, and therefore, within certain limits, individuals are to be free to pursue their preferences.

And so contemporary conservatives reject government-directed eugenic programs (in contrast to Dworkin), but they are perfectly happy with voluntary eugenics. They defend the new eugenics almost on the same ground the Left defends abortion: as a private choice to be made by individuals.

A recent roundtable discussion among several scientists and professors, published in *Prospect* magazine, wonderfully illustrates this prevalent libertarian mindset. James D. Watson, co-discoverer of DNA's structure and one of the Human Genome Project's key leaders, was asked whether he would set any limits on a form of negative eugenics, in which women have abortions for genetic reasons. "No limits. Parents should decide," he replied. When pressed by Nancy Cartwright, a professor at the London School of Economics, about creating formal mechanisms for making such decisions, he was no less adamant. The exchange went like this:

> JW: Do you want a committee of wise women like yourself telling other women what to do? You want a pseudo-consensus which, in practice, takes the decision away from individual parents.
> NC: I want a serious study and a serious public discussion.
> JW: You mean: more social science crap.

The libertarian argument that genetic engineering is acceptable so long as it is freely chosen by individuals is fraught with difficulties, however. Once genetic engineering is offered on a voluntary basis (and let's not forget that government funding has played a role in its development and thus its eventual availability), social coercion at least will come into play. Already the medical establishment pressures prospective parents to test for genetic abnormalities in the developing fetus. Conservatives should also keep in mind that liberals will demand government-funded eugenics for the economically disadvantaged. It is naive to think that if the Ivy-educated begin widely to practice eugenics, the government will not make it available to the less well-off too. Another difficulty is that individuals will not easily exercise informed, independent choices when it comes to eugenics. They will almost always need the advice of highly trained experts, and as a result their "freedom of choice" will be more form than reality. Yet because modern conservatism has been framed mainly in terms of individual freedom and opposition to Big Government, efforts to establish public standards, whether in the matter of obscenity or now eugenics, are reflexively opposed. And so it happens that though they get there by different routes, conservatives and liberals both end up supporting genetic engineering.

TECHNOPOLIS

Our inability to resist the new technologies goes beyond inadequacies in our liberal and conservative public philosophies. More broadly, our mental architecture has become furnished with technological notions and metaphors. Not long ago the editors of the *New York Times* advised: "We need to remember that the measure of a civilization is not the tools it owns but the use it makes of them." Well, of course. But one must also keep in mind that technology is more than a mere tool; it inevitably shapes our world, regardless of whether we use it wisely. Which is to say that the new eugenics seem palatable to us because technology has already profoundly affected our self-understanding. A few examples will illustrate the point.

In his elegant book *The Muse in the Machine: Computerizing the Poetry of Human Thought*, Yale computer scientist David Gelernter takes up the question of whether we will succeed in building computers that think. To assess the feasibility of *artificial* intelligence, Gelernter explains, *human* intelligence must first be defined. According to him, intelligence occurs along a spectrum: "High-focus" thought preeminently involves the manipulation of abstractions—it is, Gelernter notes, "the supreme and defining achievement of the modern mind." But abstract thought is not the whole of human thought, which also occurs at the "low-focus" level of the spectrum—found in the Bible, for example—where it is rich in symbolism, compound imagery, and counterintuitive connections. To be sure, modern, high-spectrum thought is deeper than low-spectrum thought. But it is also, Gelernter holds, narrower: "Outfitted as they are with modern minds and habits of thought, our own thinkers and philosophers are *simply unable to see* what ancient man saw."

With the full spectrum of human thought in view, one can judge scientists' hopes of constructing thinking computers—computers capable of not only high-spectrum calculations but low-spectrum visions. We will never build such computers, in Gelernter's opinion. But we will declare victory anyway, by redefining human intelligence exclusively in terms of its computer-like, high-spectrum qualities. Intelligent machines will seemingly become possible only because we have first recharacterized the human mind as a machine; computerized it and thus diminished it. In Gelernter's words: "We are not so much breaking machines to our will as rushing into their arms."

In our rush to embrace the new eugenics comparable confusions reign. We are unable to raise objections to it because our view of man is already saturated with technological categories of thought. One of the most commonly made arguments for allowing parents to increase the IQ of their children genetically is that parents already spend hundreds of thousands of dollars on education to accomplish the same goal. Thus why not allow them to reach this result more cheaply and assuredly through genetic enhancement of their children in embryo?

Formulated as such, the question answers itself. Of course parents should have this "right." But embedded in this question is a very modern, strange notion of the purpose of education: to increase IQ. Traditionally, education was not thought of in such limited, technocratic terms. To be sure, parents expected their children to learn the basics—reading, writing, math, and science. But education was thought to be much more: It was education to become a good citizen and a good man; it was about the inculcation of virtue. It was about shaping human souls, not raising test scores. If we had not lost sight of what education was properly for, if we had not come to think of it as expanding memory banks and enhancing processing speeds, we would not be open to accepting genetic engineering as a legitimate educational tool. So let's not fool ourselves: A sentiment less generous than education of the young drives the ambition to engineer smarter, cleverer beings. It is the desire for an even more complete mastery over nature.

Consider another frequently heard argument on behalf of the new eugenics. It is said that we should allow couples to select for certain desirable traits in their children, like height or blue eyes, because humans from time immemorial have practiced a crude form of eugenics every time they mated. The woman who falls for Prince Charming, or the man who courts a woman with a voluptuous figure, is said to be really seeking to produce attractive, healthy offspring. The advantage of genetic engineering is that it guarantees the result. But as James Joyce memorably put it, in *A Portrait of the Artist as a Young Man*, this view "tells you that you admired the great flanks of Venus because you felt that she would bear you burly offspring and admired her great breasts because you felt that she would give good milk to her children and yours." It is by such crude reductionism that we first come around to the scientist's dream of creating a superior race.

And strangely, we don't even recognize the danger. When the mapping of the human genome was completed last spring, President Clinton took it as proof positive that all humans were created equal after all:

> I believe one of the great truths to emerge from this triumphant expedition inside the human genome is that in genetic terms all human beings, regardless

> of race, are more than 99.9 percent the same. What that means is that modern science has confirmed what we first learned from the ancient faiths. The most important fact of life on this earth is our common humanity.

And what, one might ask, is to be made of the 0.1 percent that is not the same, that is not equal? Clinton's error, widely shared by the public, is that the principle of equality needs modern science's support. But long before modern genetics or Mendel or Darwin, the American Founders declared equality a "self-evident" truth. To them, this was a political, not a biological, truth. It was also a self-evident truth, which did not require a deep biological investigation of the human genome to be known. Rather, it could be known by all Americans, not just those trained in the abstruse science of genetics. In his appeal to modern science, Clinton may think he is bolstering our "ancient faith" in equality, but he has in fact weakened it severely. He has made a self-evident moral and political truth contingent upon what the scientists happen to come up with in their laboratories.

HUMAN BEING AND CITIZEN

No discussion of technology in America would be complete without mention of Martin Heidegger, though this is a subject to which I can hardly do justice. In his magnificent book *Reconstructing America*, James W. Ceaser describes the powerful influence Heidegger's philosophy has had on how we think about our country and technology. As Ceaser summarizes Heidegger's philosophy, the modern age represents technology's final triumph, and America technology's ultimate symbol. The importance Heidegger assigned to technology can be seen, as Ceaser shows, in that he did not analyze the different political regimes by their forms of government or principles of justice; instead, he asked which ones could spark a genuine confrontation with technology. In this reductive analysis, Nazi Germany, the Soviet Union, and America's liberal democracy were all one and the same, all embodying the catastrophe of technology. None could adequately confront technology because all were products of technology. So extreme was Heidegger's analysis that he notoriously claimed that "as for its essence, modern mechanized agriculture is the same thing as the production of dead bodies in gas chambers and extermination camps."

The falsity (and wickedness) of this claim should be obvious enough, but it illustrates the pitfalls that await any attempt to appraise technology's impact on America. There will always be the temptation to exaggerate technology's influence, to view our society, as Heidegger did, as some sort of techno-monstrosity. But there is another temptation: to view America as a grand techno-amusement park, and objections to this or that new invention as instances of Ludditism or un-Americanism. In this distorted view, as in Heidegger's, America's democratic form of government and liberal ideals as well as its Judeo-Christian heritage are just so much background noise to its true essence: Technological Innovation.

What's needed is a deeper understanding of the relation between technology and the principles of liberal democracy. To some extent one is not possible without the other. Could we have had Locke without Bacon? The former's ideals of equality and freedom without the latter's quest to master nature for the relief of man's estate? The very Constitution that secures our rights and liberties and establishes our democracy also empowers Congress "to promote the Progress of Science and useful Arts." And yet, if the political and scientific projects are inseparable from one another, they are not necessarily of equal rank and dignity.

Of all the Founding Fathers, Benjamin Franklin most embodied the spirit of scientific progress. In *The Autobiography*, he proudly recounts his invention of a new stove and enthusiastically declares this the "age of experiments." Yet one can find even in Franklin, discoverer and inventor, an awareness that science should not be a self-justifying pursuit. "There is," Franklin once counseled an aspiring student, "a prudent Moderation to be used in Studies of this Kind [natural science]." And Franklin continued: "If to attain an Eminence in that, we neglect the Knowledge and Practice of essential Duties, we deserve Reprehension. For there is no Rank in Natural Knowledge of equal Dignity and Importance with that of being a good Parent, a good Child, a good Husband or Wife, a good Neighbour or Friend, a good Subject or Citizen." Here is a promising opening for discussing and evaluating the new technologies, one that is tainted by neither Ludditism nor nihilism. It balances the benefits of science and technology against the experience of our humanity in its most fundamental relations of parent, spouse, friend, and citizen. At least it's a start.

PART II

Politics in the Genetic Age: Cloning, Stem Cells, and Beyond

Since 1997, Americans have been increasingly engaged in a debate on the genetic revolution and the American future—beginning with the announcement of Dolly the cloned sheep in 1997, and culminating in the stem cell and cloning controversies of 2001. Part II of the book attempts to chronicle that debate. The questions raised include: What will politics look like in the genetic age? What new coalitions will it create? How will it change the reigning political ideologies and parties? How will debates over human genetics transform debates over other issues—like education, health care, social security, and national defense? And is it possible to regulate the new genetics? Can one distinguish between medical therapy and enhancement, scientific experimentation and exploitation, a better human world and a new inhuman one?

SECTION A

THE CLONING/STEM CELL DEBATE, 1997–2000

When Ian Wilmut and his colleagues announced that they had cloned a newborn sheep (Dolly) from adult sheep cells, the immediate question on everyone's minds was: Is human cloning next? Was this the great technological breakthrough that would make mother-daughter twins, replacement children, and other Huxleyian nightmares an imminent possibility?

In response to the Dolly announcement, President Clinton immediately called for a voluntary moratorium on any privately funded efforts at human cloning and a temporary ban on public funding of any human cloning research. He asked his National Bioethics Advisory Commission (NBAC) to advise him on the "ethical and legal issues" involved in human cloning. In the meantime, a near universal consensus emerged that any attempt to clone human beings—"at this time," anyway—would be "irresponsible."

But within this apparent consensus there were very real disagreements: If human cloning were one day shown to be "safe"—or safe enough to try, with parental "consent"—should it be tolerated? Should it even be celebrated as a new treatment for infertility? A new reproductive right? And what about the use of cloned human embryos for medical research? Should this be tolerated? Should the federal government fund research that used such embryonic clones, which would be created simply for research and destruction?

In early 1998, the Senate debated a proposed ban on all human cloning, whether for research or reproduction. Some odd alliances emerged: Pro-choice liberals joined the biotech and pharmaceutical lobby in favor of unrestricted research. And while pro-life conservatives were the most vigorous supporters of the cloning ban, some career pro-lifers joined the pro-research faction to kill the bill. For a while, anyway, the issue lost its urgency.

Then later that year, in November 1998, James Thomson of the University of Wisconsin and John Gearhart of Johns Hopkins University announced the discovery of human embryonic stem cells—"magic cells" which they claimed might one day cure a host of dreaded diseases. A new flurry of hearings soon followed. The debate focused mostly on the moral status of embryos, which were destroyed for their stem cells, weighed against the future possibility of curing disease. Central was the question of whether this research should be eligible for public funding, which at the time it was not.

In August 2000—after another NBAC study—President Clinton and the National Institutes of Health (NIH) announced new guidelines that would circumvent the "no-research-on-embryos" ban. In short, the NIH would fund research on stem cells derived from human embryos—so long as publicly funded researchers did not do the embryo-killing and stem-cell extraction themselves; so long as the embryos were left-over from in vitro fertilization clinics; and so long as the donors of the embryos consented to their potential offspring being used in such research.

Despite some heated reaction to the Clinton guidelines, the issue of stem cell research was barely mentioned in the 2000 election. It lay in waiting as a controversy waiting to explode—and explode it did in 2001, almost immediately after George W. Bush took office.

This section includes some of the major speeches, testimony, and articles from the first wave of the cloning/stem cell debate. It includes Ian Wilmut's statement before the Senate in 1997; the reaction to Dolly; the 1998 debate over whether to ban human cloning; the discovery by James Thomson and John Gearhart of embryonic stem cells; the disagreement about the ethics of stem cell research and whether or not such work should be eligible for federal funding; and, finally, some thoughts on the election debate—or non-debate—of 2000.

TESTIMONY

U.S. SENATE

Ian Wilmut

Firstly, let me briefly summarize the experiments that my colleagues and I at the Roslin Institute and PPL Therapeutics reported in the paper that was published in *Nature* on February 27, 1997.

Individual cells were taken from the mammary gland of an adult sheep and cultured in the laboratory in conditions that allowed the cells to multiply many times. Nuclei from a few of these cells were then transferred to unfertilized sheep eggs . . . then transferred to surrogate mothers using techniques similar to those used for in vitro fertilization in humans. One of the implanted eggs developed into a normal healthy lamb that was genetically identical to the sheep from which the mammary cells were taken. Other lambs were born that had been derived from cells from sheep embryo and fetal tissue.

This is the first time that any mammal has been derived from fetal or adult cells. To understand why this is important, we need to consider some of the processes that occur during growth and development of the early embryo. Fertilization of the mammalian egg by a sperm is rapidly followed by successive cell divisions. The first few cells produced appear to be identical to each other, but by the time the sheep embryo is implanted in the womb it contains many millions of cells and several recognizable tissues. As the fetus grows the cells differentiate further so that at the end of pregnancy, the animal has hundreds of different cell types, almost all with the same original

Ian Wilmut is a biologist and joint head of the Department of Gene Expression and Development at the Roslin Institute in Edinburgh, Scotland. He was the head of the research team that successfully cloned the first mammal, Dolly the sheep. Selections from testimony before the U.S. Senate Committee on Labor and Human Resources, Subcommittee on Public Health and Safety, March 12, 1997.

genetic information as the original fertilized egg but each with a specialized function.

Scientists have tended to assume that this gradual specialization (or differentiation) was irreversible. Our previous nuclear transfer studies in which we produced lambs derived from cells from sheep embryos showed that some of the cells in the early embryo could be "reprogrammed" to develop into all the cell types present in the whole animal. Our latest work shows that cells at a much later stage of development, including some from adult animals can also be reprogrammed in the same way.

We envision using this technology in two major directions. First, in the longer term, the ability to clone larger numbers of genetically identical farm animals. Secondly, because we are now able to derive animals from cells grown in culture, we will be able to carry out much more sophisticated genetic modifications of sheep—and in the future other farm animal species—than has been possible to date.

***

Conventional genetic selection has been very successful in producing breeds of farm animals with improved productivity and as a consequence, the cost of meat, milk and eggs to the consumer has been substantially reduced. However, it takes many years for the progress that animal breeders make in their elite herds to filter down to the average commercial farmer. A limited amount of cloning within commercial breeding programs could substantially speed up this process, allowing farmers to bring up the performance of their flocks and herds closer to that of the best and enabling the livestock breeding industry to respond more quickly to changes in customer demand. . . .

We have, as yet, proven our nuclear transfer technology only in sheep. We are now just beginning research to test if it works in cattle and pigs. It will probably be at least 10 years before nuclear transfer can be used to clone cattle and pigs in commercial breeding programs.

***

The ability to derive animals from cells in culture using our nuclear transfer technology will provide a whole new range of possibilities. Genetic modification of cells in culture is relatively routine for simple organisms like bacteria and yeast and is increasingly used in mice. It has proven possible to make very precise changes in their DNA, including substitution or deletion of specific genes or the introduction of a single base pair change into the gene coding for a particular protein. We expect our results will extend this "gene targeting" capability to livestock species.

It is very expensive to produce genetically modified livestock and their use is likely to be initially confined to medical applications. PPL Therapeutics are already producing transgenic sheep and cattle that secrete large quantities of human proteins in their milk for the treatment of human diseases such as emphysema and cystic fibrosis. Others are producing transgenic pigs for use as donors of organs for transplantation to humans. These applications appear to have public support and the consumer (i.e., patient) retains a choice of whether he or she wishes to be treated with material from a transgenic animal.

Other possible uses in farm animals include the creation of animal models for human diseases, and in particular, those due to genetic defects. Several genetic diseases (e.g., cystic fibrosis) are the consequence of a single mutation in a key gene. In some circumstances, genetically modified farm animals in which the same mutation has been deliberately introduced may provide a better model for study of the treatment of such diseases than, for example, transgenic mice. We do not see meat from transgenic animals appearing on the supermarket shelves in the foreseeable future. There is currently considerable public unease in Europe about the appearance of food products made from transgenic plants and the development of transgenic farm animals for food. None of the research at Roslin is currently directed at producing transgenic farm animals for the food chain.

All experiments on animals in the UK are tightly controlled under the Animals (Scientific Procedures) Act of 1986. One of the fundamental principles of this legislation is that the potential benefits of the research must outweigh any disadvantages or harm to the animals involved. In each of the above applications, we believe that the potential benefits are large and our work presents no new challenge to public policy in the UK.

***

The potential applications that I have outlined are not, of course, the ones that have excited so much public interest. Because sheep and humans are both mammals, our results have led to widespread speculation about cloning of humans, much of it ill considered.

My own position and that of all my colleagues at the Roslin Institute and PPL Therapeutics is that [the] cloning of [a] human would be unethical.

It is also not at all clear that cloning of humans would be technically possible. The early stages of embryonic development are different among mammalian species and our methods might not transfer to those species such as pigs and humans where the embryonic genome assumes control of development at an earlier stage than in sheep.

Moreover, our own experiments to clone sheep from adult mammary cells required us to produce 277 "reconstructed" embryos. Of these, twenty-nine were implanted into recipient ewes, and only one developed into a live lamb. In previous work with cells from embryos, 3 out of 5 lambs died soon after birth and showed developmental abnormalities. Similar experiments with human[s] would be totally unacceptable.

TESTIMONY

U.S. SENATE

George J. Annas

Senator Frist, thank you for the opportunity to appear before your subcommittee to address some of the legal and ethical aspects surrounding the prospect of human cloning. I agree with President Clinton that we must "resist the temptation to replicate ourselves" and that the use of federal funds for the cloning of human beings should be prohibited. On the other hand, the contours of any broader ban on human cloning require, I believe, sufficient clarity to permit at least some research on the cellular level. This hearing provides an important opportunity to help explore and define just what makes the prospect of human cloning so disturbing to most Americans, and what steps the federal government can take to prevent the duplication of human beings without preventing vital research from proceeding.

I will make three basic points this morning: (1) the negative reaction to the prospect of human cloning by the scientific, industrial and public sectors is correct because the cloning of a human would cross a boundary that represents a difference in kind rather than in degree in human "reproduction"; (2) there are no good or sufficient reasons to clone a human; and (3) the prospect of cloning a human being provides an opportunity to establish a new regulatory framework for novel and extreme human experiments.

1. The cloning of a human would cross a natural boundary that represents a difference in kind rather than degree of human "reproduction."

George J. Annas is Edward R. Utley Professor and Chair of the Health Law Department at the Boston University Schools of Medicine and Public Health. He is the author of numerous books on health law and ethics, including *The Rights of Patients*, *Judging Medicine*, and *Health and Human Rights*. Testimony before the U.S. Senate Committee on Labor and Human Resources, Subcommittee on Public Health and Safety, March 12, 1997.

There are those who worry about threats to biodiversity by cloning animals, and even potential harm to the animals themselves. But virtually all of the reaction to the appearance of Dolly on the world stage has focused on the potential use of the new cloning technology to replicate a human being. What is so simultaneously fascinating and horrifying about this technology that produced this response? The answer is simple, if not always well-articulated: *replication of a human by cloning would radically alter the very definition of a human being by producing the world's first human with a single genetic parent.* Cloning a human is also viewed as uniquely disturbing because it is the manufacture of a person made to order, represents the potential loss of individuality, and symbolizes the scientist's unrestrained quest for mastery over nature for the sake of knowledge, power, and profits.

Human cloning has been on the public agenda before, and we should recognize the concerns that have been raised by both scientists and policy makers over the past twenty-five years. In 1972, for example, the House Subcommittee on Science, Research and Development of the Committee on Science and Astronautics asked the Science Policy Research Division of the Library of Congress to do a study on the status of genetic engineering. Among other things, that report dealt specifically with cloning and parthenogenesis as it could be applied to humans. Although the report concluded that the cloning of human beings by nuclear substitution "is not now possible," it concluded that cloning "might be considered an advanced type of genetic engineering" if combined with the introduction of highly desirable DNA to "achieve some ultimate objective of genetic engineering." The Report called for assessment and detailed knowledge, forethought and evaluation of the course of genetic developments, rather than "acceptance of the haphazard evolution of the techniques of genetic engineering [in the hope that] the issues will resolve themselves."

Six years later, in 1978, the Subcommittee on Health and the Environment of the House Committee on Interstate and Foreign Commerce held hearings on human cloning in response to the publication of David Rorvick's *The Cloning of a Man.* All of the scientists who testified assured the committee that the supposed account of the cloning of a human being was fictional, and that the techniques described in the book could not work. One scientist testified that he hoped that by showing that the report was false it would also become apparent that the issue of human cloning itself "is a false one, that the apprehensions people have about cloning of human beings are totally unfounded." The major point the scientists wanted to make, however, was that they didn't want any regulations that might affect their research. In the words of one, "There is no need for any form of regulatory legislation, and it could only in the long run have a harmful effect."

Congressional discussion of human cloning was interrupted by the birth of Baby Louise Brown, the world's first IVF baby, in 1978. The ability to conceive a child in a laboratory not only added a new way (in addition to artificial insemination) for humans to reproduce without sex, but also made it possible for the first time for a woman to gestate and give birth to a child to whom she had no genetic relationship. Since 1978, a child can have at least five parents: a genetic and rearing father, and a genetic, gestational, and rearing mother. We pride ourselves as having adapted to this brave new biological world, but in fact we have yet to develop reasonable and enforceable rules for even so elementary a question as who among these five possible parents the law should recognize as those with rights and obligations to the child. Many other problems, including embryo storage and disposition, posthumous use of gametes, and information available to the child also remain unresolved.

IVF represents a striking technological approach to infertility; nonetheless the child is still conceived by the union of an egg and sperm from two separate human beings of the opposite sex. Even though no change in the genetics and biology of embryo creation and growth is at stake in IVF, society continues to wrestle with fundamental issues involving this method of reproduction 20 years after its introduction. Viewing IVF as a precedent for human cloning misses the point. Over the past two decades many ethicists have been accused of "crying wolf" when new medical and scientific technologies have been introduced. This may have been the case in some instances, but not here. This change in kind in the fundamental way in which humans can "reproduce" represents such a challenge to human dignity and the potential devaluation of human life (even comparing the "original" to the "copy" in terms of which is to be more valued) that even the search for an analogy has come up empty handed.

Cloning is replication, not reproduction, and represents a difference in kind not in degree in the manner in which human beings reproduce. Thus, although the constitutional right not to reproduce would seem to apply with equal force to a right not to replicate, to the extent that there is a constitutional right to reproduce (if one is able to), it seems unlikely that existing privacy or liberty doctrine would extend this right to replication by cloning.

2. *There are no good or sufficient reasons to clone a human.*

When the President's Bioethics Commission reported on genetic engineering in 1982 in their report entitled *Splicing Life*, human cloning rated only a short paragraph in a footnote. The paragraph concluded: "The technology to clone a human does not—and may never—exist. Moreover, the critical nongenetic influences on development make it difficult to imagine producing a human clone who would act or appear 'identical'." The NIH Human Embryo Research panel that reported on human embryo research in September 1994 also devoted only a single footnote to this type of cloning. "Popular notions of cloning derive from science fiction books and films that have more to do with cultural fantasies than actual scientific experiments." Both of these expert panels were wrong to disregard lessons from our literary heritage on this topic, thereby attempting to sever science from its cultural context.

Literary treatments of cloning help inform us that applying this technology to humans is too dangerous to human life and values. The reporter who described Dr. Ian Wilmut as "Dolly's laboratory father" couldn't have conjured up images of Mary Shelley's Frankenstein better if he had tried. Frankenstein was also his creature's father/god; the creature telling him: "I ought to be thy Adam." Like Dolly, the "spark of life" was infused into the creature by an electric current. Unlike Dolly, the creature was created as a fully grown adult (not a cloning possibility, but what many Americans fantasize and fear), and wanted more than creaturehood: he wanted a mate of his "own kind" with whom to live, and reproduce. Frankenstein reluctantly agreed to manufacture such a mate if the creature agrees to leave humankind alone, but in the end, viciously destroyed the female creature-mate, concluding that he has no right to inflict the children of this pair, "a race of devils," upon "everlasting generations." Frankenstein ultimately recognized his responsibilities to humanity, and Shelley's great novel explores virtually all the noncommercial elements of today's cloning debate.

The naming of the world's first cloned mammal also has great significance. The sole survivor of 277 cloned embryos (or "fused couplets"), the clone could have been named after its sequence number in this group (e.g., C-137), but this would have only emphasized its character as a produced product. In stark contrast, the name Dolly (provided for the public and not used in the scientific report in *Nature*) suggests an individual, a human or at least a pet. Even at the manufactured level a "doll" is something that produces great joy in our children and is itself harmless. Victor Frankenstein, of course, never named his creature, thereby repudiating any parental responsibility. The creature himself evolved into a monster when it was rejected not only by Frankenstein, but by society as well. Naming the world's first mammal-clone Dolly is meant to distance her from the Frankenstein myth both by making her appear as something she is not, and by assuming parental obligations toward her.

Unlike Shelley's, Aldous Huxley's Brave New World future in which all humans are created by cloning through embryo splitting and conditioned to join one of five worker groups, was always unlikely. There are much more efficient ways of creating killers or terrorists (or even workers) than through cloning—physical and psychological conditioning can turn teenagers into terrorists in a matter of months, rather than waiting some eighteen to twenty years for the clones to grow up and be trained themselves. Cloning has no real military or paramilitary uses. Even Hitler's clone would himself likely be quite a different person because he would grow up in a radically altered world environment.

It has been suggested, however, that there might be good reasons to clone a human. Perhaps most compelling is cloning a dying child if this is what the grieving parents want. But this should not be permitted. Not only does this encourage the parents to produce one child in the image of another, it also encourages all of us to view children as interchangeable commodities, because cloning is so different from human reproduction. When a child is cloned, it is not the parents that are being replicated (or are "reproducing") but the child. No one should have such dominion over a child (even a dead or dying child) as to be permitted to use its genes to create the child's child. Humans have a basic right not to reproduce, and human reproduction (even replication) is not like reproducing farm animals, or even pets. Ethical human reproduction properly requires the voluntary participation of the genetic parents. Such voluntary participation is not possible for a young child. Related human rights and dignity would also prohibit using cloned children as organ sources for their father/mother original. Nor is there any "right to be cloned" that an adult might possess that is triggered by marriage to someone with whom the adult cannot reproduce.

Any attempt to clone a human being should also be prohibited by basic ethical principles that prohibit putting human subjects at significant risk without their informed consent. Dolly's birth was a one in 277 embryo chance. The birth of a human from cloning might be technologically possible, but we could only discover this by unethically subjecting the planned child to the risk of serious genetic or physical injury, and subjecting a planned child to this type of risk could literally never be justified. Because we will likely never be able to protect the human subject of cloning research from serious harm, the basic ethical rules of human experimentation prohibit us from ever using it on humans.

3. Developing a regulatory framework for human cloning.

What should we do to prevent Dolly technology from being used to manufacture duplicate humans? We have three basic models for scientific/medical policy-making in the U.S.: the market, professional standards, and legislation. We tend to worship the first, distrust the second, and disdain the third. Nonetheless, the prospect of human cloning requires more deliberation about social and moral issues than either the market or science can provide. The market has no morality, and if we believe important values including issues of human rights and human dignity are at stake, we cannot leave cloning decisions to the market. The Biotechnology Industry Organization in the U.S. has already taken the commendable position that human cloning should be prohibited by law. Science often pretends to be value-free, but in fact follows its own imperatives, and either out of ignorance or self-interest assumes that others are making the policy decisions about whether or how to apply the fruits of their labors. We disdain government involvement in reproductive medicine. But cloning is different in kind, and only government has the authority to restrain science and technology until its social and moral implications are adequately examined.

We have a number of options. The first is for Congress to simply ban the use of human cloning. Cloning for replication can (and should) be confined to nonhuman life. We need not, however, prohibit all possible research at the cellular level. For example, to the extent that scientists can make a compelling case for use of cloning technology on the cellular level for research on processes such as cell differentiation and senescence, and so long as any and all attempts to implant a resulting embryo into a human or other animal, or to continue cell division beyond a 14-day period are prohibited, use of human cells for research could be permitted. Anyone proposing such research, however, should have the burden of proving that the research is vital, cannot be conducted any other way, and is unlikely to produce harm to society.

The prospect of human cloning also provides Congress with the opportunity to go beyond ad hoc bans on procedures and funding, and the periodic appointment of blue ribbon committees, and to establish a Human Experimentation Agency with both rule-making and adjudicatory authority in the area of human experimentation. Such an agency could both promulgate rules governing human research and review and approve or disapprove research proposals in areas such as human cloning which local IRBs are simply incapable of providing meaningful reviews. The President's Bioethics panel is important and useful as a forum for discussion and possible policy development. But we have had such panels before, and it is time to move beyond discussion to meaningful regulation in areas like cloning where there is already a societal consensus.

One of the most important procedural steps a Federal Human Experimentation Agency should take is to put the burden of proof on those who propose to do extreme and novel experiments, such as cloning, that cross recognized boundaries and call deeply held societal values into question. Thus, cloning proponents should have to prove that there is a compelling reason to approve research on it. I think the Canadian Royal Commission on New Reproductive Technologies quite properly concluded that both cloning and embryo splitting have "no foreseeable ethically acceptable application to the human situation" and therefore should not be done. We need an effective mechanism to ensure that it is not.

THE WISDOM OF REPUGNANCE

Leon R. Kass

Our habit of delighting in news of scientific and technological breakthroughs has been sorely challenged by the birth announcement of a sheep named Dolly. Though Dolly shares with previous sheep the "softest clothing, woolly, bright," William Blake's question, "Little Lamb, who made thee?" has for her a radically different answer: Dolly was, quite literally, made. She is the work not of nature or nature's God but of man, an Englishman, Ian Wilmut, and his fellow scientists. What's more, Dolly came into being not only asexually—ironically, just like "He [who] calls Himself a Lamb"—but also as the genetically identical copy (and the perfect incarnation of the form or blueprint) of a mature ewe, of whom she is a clone. This long-awaited yet not quite expected success in cloning a mammal raised immediately the prospect—and the specter—of cloning human beings: "I a child and Thou a lamb," despite our differences, have always been equal candidates for creative making, only now, by means of cloning, we may both spring from the hand of man playing at being God.

***

[C]loning personifies our desire fully to control the future, while being subject to no controls ourselves. Enchanted and enslaved by the glamour of technology, we have lost our awe and wonder before the deep mysteries of nature and of life. We cheerfully take our own beginnings in our hands and, like the last man, we blink.

Selections from "The Wisdom of Repugnance," *The New Republic,* June 2, 1997.

Part of the blame for our complacency lies, sadly, with the field of bioethics itself, and its claim to expertise in these moral matters. Bioethics was founded by people who understood that the new biology touched and threatened the deepest matters of our humanity: bodily integrity, identity and individuality, lineage and kinship, freedom and self-command, eros and aspiration, and the relations and strivings of body and soul. With its capture by analytic philosophy, however, and its inevitable routinization and professionalization, the field has by and large come to content itself with analyzing moral arguments, reacting to new technological developments and taking on emerging issues of public policy, all performed with a naive faith that the evils we fear can all be avoided by compassion, regulation and a respect for autonomy. Bioethics has made some major contributions in the protection of human subjects and in other areas where personal freedom is threatened; but its practitioners, with few exceptions, have turned the big human questions into pretty thin gruel.

One reason for this is that the piecemeal formation of public policy tends to grind down large questions of morals into small questions of procedure. Many of the country's leading bioethicists have served on national commissions or state task forces and advisory boards, where, understandably, they have found utilitarianism to be the only ethical vocabulary acceptable to all participants in discussing issues of law, regulation and public policy. As many of these commissions have been either officially under the aegis of NIH or the Health and Human Services Department, or otherwise dominated by powerful voices for scientific progress, the ethicists have for the most part been content, after some "values clarification" and wringing of hands, to pronounce their blessings upon the inevitable. Indeed, it is the bioethicists, not the scientists, who are now the most articulate defenders of human cloning: the two witnesses testifying before the National Bioethics Advisory Commission in favor of cloning human beings were bioethicists, eager to rebut what they regard as the irrational concerns of those of us in opposition. One wonders whether this commission, constituted like the previous commissions, can tear itself sufficiently free from the accommodationist pattern of rubber-stamping all technical innovation, in the mistaken belief that all other goods must bow down before the gods of better health and scientific advance.

If it is to do so, the commission must first persuade itself, as we all should persuade ourselves, not to be complacent about what is at issue here. Human cloning, though it is in some respects continuous with previous reproductive technologies, also represents something radically new, in itself and in its easily foreseeable consequences. The stakes are very high indeed. I exaggerate, but in the direction of the truth, when I insist that we are faced with having to decide nothing less than whether human procreation is going to remain human, whether children are going to be made rather than begotten, whether it is a good thing, humanly speaking, to say yes in principle to the road which leads (at best) to the dehumanized rationality of *Brave New World*. This is not business as usual, to be fretted about for a while but finally to be given our seal of approval. We must rise to the occasion and make our judgments as if the future of our humanity hangs in the balance. For so it does.

CLONING HUMAN BEINGS

REPORT AND RECOMMENDATIONS

National Bioethics Advisory Commission

The idea that humans might someday be cloned—created from a single somatic cell without sexual reproduction—moved further away from science fiction and closer to a genuine scientific possibility on February 23, 1997. On that date, *The Observer* broke the news that Ian Wilmut, a Scottish scientist, and his colleagues at the Roslin Institute were about to announce the successful cloning of a sheep by a new technique which had never before been fully successful in mammals. The technique involved transplanting the genetic material of an adult sheep, apparently obtained from a differentiated somatic cell, into an egg from which the nucleus had been removed. The resulting birth of the sheep, named Dolly, on July 5, 1996, was different from prior attempts to create identical offspring since Dolly contained the genetic material of only one parent, and was, therefore, a "delayed" genetic twin of a single adult sheep.

***

Within days of the published report of Dolly, President Clinton instituted a ban on federal funding related to attempts to clone human beings in this manner. In addition, the President asked the recently appointed National Bioethics Advisory Commission (NBAC) to address within ninety days the ethical and legal issues that surround the subject of cloning human beings.

The National Bioethics Advisory Commission (NBAC) was created by President Bill Clinton in October 1995 to provide advice and recommendations on issues related to human biological or behavioral research. Its major reports were on human cloning, stem cells, and the ethics of experimentation with human subjects. Selections from "Cloning Human Beings: Report and Recommendations of the National Bioethics Advisory Commission," June 1997.

This provided a welcome opportunity for initiating a thoughtful analysis of the many dimensions of the issue, including a careful consideration of the potential risks and benefits. It also presented an occasion to review the current legal status of cloning and the potential constitutional challenges that might be raised if new legislation were enacted to restrict the creation of a child through somatic cell nuclear transfer cloning.

The Commission began its discussions fully recognizing that any effort in humans to transfer a somatic cell nucleus into an enucleated egg involves the creation of an embryo, with the apparent potential to be implanted in utero and developed to term. Ethical concerns surrounding issues of embryo research have recently received extensive analysis and deliberation in the United States. Indeed, federal funding for human embryo research is severely restricted, although there are few restrictions on human embryo research carried out in the private sector. Thus, under current law, the use of somatic cell nuclear transfer to create an embryo solely for research purposes is already restricted in cases involving federal funds. There are, however, no current federal regulations on the use of private funds for this purpose.

The unique prospect, vividly raised by Dolly, is the creation of a new individual genetically identical to an existing (or previously existing) person—a "delayed" genetic twin. This prospect has been the source of the overwhelming public concern about such cloning. While the creation of embryos for research purposes alone always raises serious ethical questions, the use of somatic cell nuclear transfer to create embryos raises no new issues in this respect. The unique and distinctive ethical issues raised by the use of somatic cell nuclear transfer to create children relate to, for example, serious safety concerns, individuality, family integrity, and treating children as objects. Consequently, the Commission focused its attention on the use of such techniques for the purpose of creating an embryo which would then be implanted in a woman's uterus and brought to term. It also expanded its analysis of this particular issue to encompass activities in both the public and private sector.

In its deliberations, NBAC reviewed the scientific developments which preceded the Roslin announcement, as well as those likely to follow in its path. It also considered the many moral concerns raised by the possibility that this technique could be used to clone human beings. Much of the initial reaction to this possibility was negative. Careful assessment of that response revealed fears about harms to the children who may be created in this manner, particularly psychological harms associated with a possibly diminished sense of individuality and personal autonomy. Others expressed concern about a degradation in the quality of parenting and family life.

In addition to concerns about specific harms to children, people have frequently expressed fears that the widespread practice of somatic cell nuclear transfer cloning would undermine important social values by opening the door to a form of eugenics or by tempting some to manipulate others as if they were objects instead of persons. Arrayed against these concerns are other important social values, such as protecting the widest possible sphere of personal choice, particularly in matters pertaining to procreation and child rearing, maintaining privacy and the freedom of scientific inquiry, and encouraging the possible development of new biomedical breakthroughs.

To arrive at its recommendations concerning the use of somatic cell nuclear transfer techniques to create children, NBAC also examined longstanding religious traditions that guide many citizens' responses to new technologies and found that religious positions on human cloning are pluralistic in their premises, modes of argument, and conclusions. Some religious thinkers argue that the use of somatic cell nuclear transfer cloning to create a child would be intrinsically immoral and thus could never be morally justified. Other religious thinkers contend that human cloning to create a child could be morally justified under some circumstances, but hold that it should be strictly regulated in order to prevent abuses.

The public policies recommended with respect to the creation of a child using somatic cell nuclear transfer reflect the Commission's best judgments about both the ethics of attempting such an experiment and its view of traditions regarding limitations on individual actions in the name of the common good. At present, the use of this technique to create a child would be a premature experiment that would expose the fetus and the developing child to unacceptable risks. This in itself might be sufficient to justify a prohibition on cloning human beings at this time, even if such efforts were to be characterized as the exercise of a fundamental right to attempt to procreate.

Beyond the issue of the safety of the procedure, however, NBAC found that concerns relating to the potential psychological harms to children and effects on the moral, religious, and cultural values of society merited further reflection and deliberation. Whether upon such further deliberation our nation will conclude that the use of cloning techniques to create children should be allowed or permanently banned is, for the moment, an open question. Time is an ally in this regard, allowing for the accrual of further data from animal experimentation, enabling an assessment of the prospective safety and efficacy of the procedure in humans, as well as granting a period of fuller national debate on ethical and social concerns. The Commission therefore concluded that there should be imposed a period of time in which no attempt is made to create a child using somatic cell nuclear transfer.

TESTIMONY

U.S. Senate

Edmund D. Pellegrino

President Clinton and his National Commission on Bioethics have recently announced that human cloning . . . is morally wrong and should be banned because of its potential for grave physical and psychosocial harm both to the child and to fundamental family and social values American society cherishes. These harms they believe outweigh the putative benefits of human cloning in treating and preventing disease.

Their reasons for recommending a ban are morally sound as far as they go. They are consistent with the moral principles of nonmalfeasance and respect for persons. Unfortunately, the President and his Commission propose a temporary ban of five years. This undercuts the moral probity of their recommendation for two reasons. First, it suggests that something inherently wrong can be in time made right; and second, it begs the question of the moral wrong of human embryo experimentation, which is the first and essential step in any cloning of human beings.

I will confine my remarks to these two moral inconsistencies, since they are likely to be at the heart of any discussion of the President's proposed Cloning Prohibition Act of 1997.

The President and his Commission propose a temporary five year ban on cloning to gain time to study the nature and degrees of potential harms of cloning. They deem the procedure "not safe at this point" but would permit

Edmund D. Pellegrino is Professor Emeritus of Medicine and Medical Ethics at Georgetown University. He is the author of numerous books and articles in medical science, philosophy, and ethics and founding editor of the *Journal of Medicine and Philosophy*. Selections from testimony before the U.S. Senate Committee on Labor and Human Resources, Subcommittee on Public Health and Safety, June 17, 1997.

cloning of embryos so long as there is no intent to implant them in a woman's womb.

This assertion compounds two moral errors. First it assumes that something grievously wrong can become morally right by extended discussion "before this technology can be used." This converts the grievous harms they now see into matters of preference and attitude, making moral truth the creature of public opinion and plebiscite. The moral harms of cloning are inherent in the concept itself, and in the fact that obtaining further information about harms and hazards depends upon the deliberate manufacture, manipulation, and destruction of human embryos.

Human embryos are members of the human species in its earliest and most vulnerable stages. When they are manufactured to answer questions of harm they themselves are harmed and thus become experimental subjects created to serve the interests of others. The President's legislation prohibits manufacturing of embryos with intent to implant but it does not forbid making embryos. Implicitly it invites their manufacture as a means of gathering information in the interim of the ban. Thousands of embryos will be deliberately created, manipulated, and destroyed. We must not forget that it took 277 attempts to make one sheep.

Embryo research recommended three years ago by a special NIH panel is presently under presidential ban. Yet the Commission's report tacitly accepts the NIH panel's recommendation and gives it new life by suggesting that embryo research will be needed to resolve unanswered questions. This is a logically and morally inconsistent argument. To be consistent a permanent ban should be placed on human cloning because it depends on a first step which is itself morally indefensible.

Cloning of DNA molecules or of individual human cells and tissues not taken from manufactured, cloned embryos is morally licit. The information obtained can be useful in diagnosing, treating, and preventing human genetic disorders without resort to cloning whole human beings.

A permanent ban on human cloning, while morally correct, will not prevent uses and abuses of this technique. A few weeks ago the *New York Times* reported a person named Randolph Wicker who has decided to seek a scientist to help him clone himself. Human narcissism and commercial advantage suggest that humans will be cloned. Legislation cannot assure that citizens will act morally. What it can do is restrain an inherently immoral practice, refuse social legitimation, and provide legal recourse for children harmed by the narcissistic predilections of those who seek to create new beings in their own images.

SECOND THOUGHTS ON CLONING

Laurence H. Tribe

Some years ago, long before human cloning became a near-term prospect, I was among those who urged that human cloning be assessed not simply in terms of concrete costs and benefits, but in terms of what the technology might do to the very meaning of human reproduction, child rearing and individuality. I leaned toward prohibition as the safest course.

Today, with the prospect of a renewed push for sweeping prohibition rather than mere regulation, I am inclined to say, "Not so fast."

When scientists announced in February that they had created a clone of an adult sheep—a genetically identical copy named Dolly, created in the laboratory from a single cell of the "parent"—ethicists, theologians and others passionately debated the pros and cons of trying to clone a human being.

People spoke of the plight of infertile couples; the grief of someone who has lost a child whose biological "rebirth" might offer solace; the prospect of using cloning to generate donors for tissues and organs; the possibility of creating genetically enhanced clones with a particular talent or a resistance to some dread disease.

But others saw a nightmarish and decidedly unnatural perversion of human reproduction. California enacted a ban on human cloning, and the President's National Bioethics Advisory Commission recommended making the ban nationwide.

Laurence H. Tribe is a professor of constitutional law at Harvard University. He has written widely on abortion, civil liberties, federalism, privacy, the separation of church and state, and the Supreme Court, and his many books include *Constitutional Choices* and *Abortion: The Clash of Absolutes*.
This essay originally appeared in the *New York Times* on December 5, 1997.

That initial debate has cooled, however, and many in the scientific field now seem to be wondering what all the fuss was about.

They are asking whether human cloning isn't just an incremental step beyond what we are already doing with artificial insemination, in vitro fertilization, fertility enhancing drugs and genetic manipulation. That casual attitude is sure to give way before long to yet another wave of prohibitionist outrage—a wave that I no longer feel comfortable riding.

I certainly don't subscribe to the view that whatever technology permits us to do we ought to do. Nor do I subscribe to the view that the Constitution necessarily guarantees every individual the right to reproduce through whatever means become technically possible.

Rather, my concern is that the very decision to use the law to condemn, and then outlaw, patterns of human reproduction—especially by invoking vague notions of what is "natural"—is at least as dangerous as the technologies such a decision might be used to control.

Human cloning has been condemned by some of its most articulate detractors as the ultimate embodiment of the sexual revolution, severing sex from the creation of babies and treating gender and sexuality as socially constructed.

But to ban cloning as the technological apotheosis of what some see as culturally distressing trends may, in the end, lend credence to strikingly similar objections to surrogate motherhood or gay marriage and gay adoption.

Equally scary, when appeals to the natural, or to the divinely ordained, lead to the criminalization of some method for creating human babies, we must come to terms with the inevitable: the prohibition will not be airtight.

Just as was true of bans on abortion and on sex outside marriage, bans on human cloning are bound to be hard to enforce. And that, in turn, requires us to think in terms of a class of potential outcasts—people whose very existence society will have chosen to label as a misfortune and, in essence, to condemn.

One need only think of the long struggle to overcome the stigma of "illegitimacy" for the children of unmarried parents. How much worse might be the plight of being judged morally incomplete by virtue of one's man-made origin?

There are some black markets (in narcotic drugs, for instance) that may be worth risking when the evils of legalization would be even worse. But when the contraband we are talking of creating takes the form of human beings, the stakes become enormous.

There are few evils as grave as that of creating a caste system, one in which an entire category of persons, while perhaps not labeled untouchable, is marginalized as not fully human.

And even if one could enforce a ban on cloning, or at least insure that clones would not be a marginalized caste, the social costs of prohibition could still be high. For the arguments supporting an ironclad prohibition of cloning are most likely to rest on, and reinforce, the notion that it is unnatural and intrinsically wrong to sever the conventional links between heterosexual unions sanctified by tradition and the creation and upbringing of new life.

The entrenchment of that notion cannot be a welcome thing for lesbians, gay men and perhaps others with unconventional ways of linking erotic attachment, romantic commitment, genetic replication, gestational mothering and the joys and responsibilities of child rearing.

And, from the perspective of the wider community, straight no less than gay, a society that bans acts of human creation for no better reason than that their particular form defies nature and tradition is a society that risks cutting itself off from vital experimentation, thus losing a significant part of its capacity to grow. If human cloning is to be banned, then, the reasons had better be far more compelling than any thus far advanced.

OF HEADLESS MICE . . . AND MEN

Charles Krauthammer

Last year Dolly the cloned sheep was received with wonder, titters and some vague apprehension. Last week the announcement by a Chicago physicist that he is assembling a team to produce the first human clone occasioned yet another wave of Brave New World anxiety. But the scariest news of all—and largely overlooked—comes from two obscure labs, at the University of Texas and at the University of Bath. During the past four years, one group created headless mice; the other, headless tadpoles.

For sheer Frankenstein wattage, the purposeful creation of these animal monsters has no equal. Take the mice. Researchers found the gene that tells the embryo to produce the head. They deleted it. They did this in a thousand mice embryos, four of which were born. I use the term loosely. Having no way to breathe, the mice died instantly.

Why then create them? The Texas researchers want to learn how genes determine embryo development. But you don't have to be a genius to see the true utility of manufacturing headless creatures: for their organs—fully formed, perfectly useful, ripe for plundering.

Why should you be panicked? Because humans are next. "It would almost certainly be possible to produce human bodies without a forebrain," Princeton biologist Lee Silver told the London *Sunday Times*. "These human bodies without any semblance of consciousness would not be considered persons, and thus it would be perfectly legal to keep them 'alive' as a future source of organs."

Charles Krauthammer is a syndicated columnist for the *Washington Post*, a contributor to *Time* magazine, and a contributing editor to the *Weekly Standard*. He served as a science adviser to President Jimmy Carter and won the Pulitzer Prize for commentary in 1987. This essay first appeared in *Time* magazine on January 19, 1998.

"Alive." Never have a pair of quotation marks loomed so ominously. Take the mouse-frog technology, apply it to humans, combine it with cloning, and you are become a god: with a single cell taken from, say, your finger, you produce a headless replica of yourself, a mutant twin, arguably lifeless, that becomes your own personal, precisely tissue-matched organ farm.

There are, of course, technical hurdles along the way. Suppressing the equivalent "head" gene in man. Incubating tiny infant organs to grow into larger ones that adults could use. And creating artificial wombs (as per Aldous Huxley), given that it might be difficult to recruit sane women to carry headless fetuses to their birth/death.

It won't be long, however, before these technical barriers are breached. The ethical barriers are already cracking. Lewis Wolpert, professor of biology at University College, London, finds producing headless humans "personally distasteful" but, given the shortage of organs, does not think distaste is sufficient reason not to go ahead with something that would save lives. And Professor Silver not only sees "nothing wrong, philosophically or rationally," with producing headless humans for organ harvesting; he wants to convince a skeptical public that it is perfectly O.K.

When prominent scientists are prepared to acquiesce in—or indeed encourage—the deliberate creation of deformed and dying quasi-human life, you know we are facing a bioethical abyss. Human beings are ends, not means. There is no grosser corruption of biotechnology than creating a human mutant and disemboweling it at our pleasure for spare parts.

The prospect of headless human clones should put the whole debate about "normal" cloning in a new light. Normal cloning is less a treatment for infertility than a treatment for vanity. It is a way to produce an exact genetic replica of yourself that will walk the earth years after you're gone.

But there is a problem with a clone. It is not really you. It is but a twin, a perfect John Doe Jr., but still a junior. With its own independent consciousness, it is, alas, just a facsimile of you.

The headless clone solves the facsimile problem. It is a gateway to the ultimate vanity: immortality. If you create a real clone, you cannot transfer your consciousness into it to truly live on. But if you create a headless clone of just your body, you have created a ready source of replacement parts to keep you—your consciousness—going indefinitely.

Which is why one form of cloning will inevitably lead to the other. Cloning is the technology of narcissism, and nothing satisfies narcissism like immortality. Headlessness will be cloning's crowning achievement.

The time to put a stop to this is now. Dolly moved President Clinton to create a commission that recommended a temporary ban on human cloning. But with physicist Richard Seed threatening to clone humans, and with headless animals already here, we are past the time for toothless commissions and meaningless bans.

Clinton banned federal funding of human-cloning research, of which there is none anyway. He then proposed a five-year ban on cloning. This is not enough. Congress should ban human cloning now. Totally. And regarding one particular form, it should be draconian: the deliberate creation of headless humans must be made a crime, indeed a capital crime. If we flinch in the face of this high-tech barbarity, we'll deserve to live in the hell it heralds.

WHO'S AFRAID OF HUMAN CLONING?

James K. Glassman

Ever since Galileo ran afoul of the Inquisition by propounding the idea that the planets move around the sun, the powers that be have reflexively tried to halt the progress of science into areas unknown and uncomfortable.

The latest effort is underway right now in Congress, where a bill to ban human cloning is moving swiftly. Introduced by Republicans Bill Frist and Kit Bond, it is a colossal mistake—a triumph of superstition, government coercion, self-righteousness and fear over good sense, health, family values and confidence in the future.

The bill makes it a felony, punishable by a prison term of up to 10 years, for scientists to conduct research using "somatic nuclear cell transfer"—the same dazzling procedure that created the cloned sheep Dolly—to produce a human clone or a cloned human embryo.

Nearly all bans on research prove to be temporary; science eventually trumps politics. But in the meantime, as Gregory Benford of the University of California writes in *Reason* magazine, there is a high cost in "lives lost because the resultant technology arrives too late for some patients."

Also, with scientific change coming faster, the cloning ban sets a dangerous precedent, validating a knee-jerk response to technology. Even Frist acknowledged last week that "we're banning something before it's fully understood." In my book, that's an excellent definition of ignorance.

James K. Glassman is a resident fellow at the American Enterprise Institute and host of TechCentralStation.com, a Web site on politics and technology. He is the co-author of *Dow 36,000* (with Kevin A. Hassett) and a regular columnist for the *International Herald Tribune*. This essay originally appeared in the *Washington Post* on February 10, 1998.

Frist and Bond aren't alone. President Clinton, who banned federal cloning research a year ago, calls the procedure "morally unacceptable."

Majority Leader Dick Armey—whose letterhead, ironically, carries the slogan "Freedom Works"—is the most egregious anti-cloning zealot of all. "Congress should enact a permanent ban on human cloning," he said recently, "to keep this frightening idea the province of the mad scientists of science fiction."

In fact, scientists who are interested in cloning aren't mad at all. They want either to help sick people live longer, healthier lives or to help infertile people have babies.

A year ago, Ian Wilmut, a Scottish embryologist, created a little lamb by taking a normal sheep egg cell, removing the nucleus (the cell structure that contains the genes) and placing a cell from the mammary tissue of an adult sheep alongside it.

An electrical charge then fused the two cells into one. Remarkably, the new cell reverted to its embryonic state and started dividing. It developed into a fetus and then a newborn lamb that was identical, in its genetic makeup, to the adult lamb that contributed the mammary cell. Anti-cloning hysterics say that 276 embryos were destroyed to create Dolly, but Gina Kolata, the *New York Times* reporter who broke the story, and the author of the new book *Clone*, told me this was "a gross distortion."

Here's what happened, according to the paper in the journal *Nature*: 277 eggs underwent fusion but only 29 survived for as much as five days. These clumps of cells were then transferred into 13 female sheep. Three months later, ultrasound tests found that only one fetus (Dolly) survived. "There were no monsters, no miscarriages, nothing," says Kolata. Immoral?

Today, she adds, the technology is so advanced that even these inefficiencies are an anachronism.

The technique will make all kinds of medical research easier and faster. For instance, by studying the way cells revert and divide, we may be able to find cures for Alzheimer's and Parkinson's diseases and to grow new organs to replace damaged ones.

A report by NIH says that somatic nuclear transfer "might also be used in the future to create skin grafts for people who are severely burned" and to generate new cells to treat liver damage, leukemia, sickle-cell and heart disease.

In addition, the technology could be used in gene therapy, which can replace a patient's defective DNA (which will cause disease) with a normal gene, perhaps very early in life, and prevent AIDS and cancer.

Some of this can be accomplished with cloning techniques that manage to get around the restrictions in the Frist-Bond bill, but why have limits at all? Because such research is somehow "unnatural"? Because people should have babies only by "natural" means?

"If humans have a right to reproduce, what right does society have to limit the means?" wrote Nathan Myhrvold, chief technology officer of Microsoft Corp., last year in *Slate*, the electronic magazine.

Remember the uproar over "test-tube" babies just two decades ago? With in-vitro fertilization, the sperm and egg are combined in the lab and surgically implanted in the womb. "To date," writes Myhrvold, "nearly 30,000 such babies have been born in the United States alone. Many would-be parents have been made happy. Who has been harmed?"

The only difference between these babies and clones is that the combined DNA of the egg and sperm is replaced by the DNA of only one adult donor, inserted into a "hollowed-out" egg. Now politicians want to pass a law that says that one combination is fine but the other is criminal.

It's astounding that groups that represent Americans with serious diseases aren't outraged by what Congress is doing—not to mention infertile couples and scientists and physicians who value their freedom of inquiry.

And it's astounding that Republicans—whose political hopes lie not with reactionaries who fear the future but with people (comfortable with high technology and the Internet) who embrace it—should be in the vanguard of the cloning ban. Dumb and dumber.

SPEECH

U.S. SENATE

Senator Tom Harkin

Now, there has been a lot of, I think, undue, inflammatory kinds of statements and comments made about this cloning research. It seems odd to me that on something that has so much potential to alleviate human suffering and which is also, I will be frank to admit, fraught with perils of ethics and bioethics—it seems odd to me that a bill of that nature would be rushed so soon to the floor of the Senate. It seems to me that this is the kind of bill that ought to go through a lengthy and involved hearing process, to bring in the best minds, ethicists, physicians, doctors, researchers, those involved in gene therapy, those who have been involved in cloning research in the past, to hear their views on this. And then out of this, perhaps we can develop a more reasoned, logical, bipartisan approach on the issue of cloning research.

So I have to ask, what is this so-called rush? Why bring it out on the floor like this without the proper kind of hearings, because there is a hidden political agenda? Is this to inflame fears among people? Well, I hope not. To take away that apprehension, I think the best thing would be to refer this to committee and have hearings on it. I serve on the Labor, Health and Human Services Committee, and I would assume that committee would be the proper one to have the hearings, at least some of them, plus those on the House side. So I want to speak about it in that context.

Mr. President, each year, too many of our loved ones suffer terribly. They are taken away from us by diseases like cancer, heart disease and Alzheimer's.

Senator Tom Harkin (D) is a member of the U.S. Senate from Iowa. He was elected to the Senate in 1984, where he was reelected in 1990 and 1996. He has been one of the most outspoken advocates for advancing medical research and was a leading opponent of the effort to ban human cloning in 1998. Selections from *Congressional Record*, February 9, 1998.

For many years, I have worked hard to expand research into finding cures and preventative measures and improve treatments for the many conditions that rob us of our health. Over the last several years, there have been major breakthroughs in medical research. We need to make sure that our world-class scientists continue to build on this progress, but that we also say to young people who are in college today, maybe even in high school, who are thinking of pursuing research careers, that we welcome their inquisitiveness, we welcome their experimentation, we want there to be no bounds put on their inquiries by a rush to judgment by the Congress of the United States, which is ill-equipped to make such a judgment. I think our actions here send a very chilling message to young people, who want to go into biomedical research, that somehow there is going to be the heavy hand of "Big Brother" Government overlooking their research, telling them you can do this but not that, or you can go no further than that, or you can ask this question, but you can't ask that question. I think this bill that we have, again, pushed before us in this rush, can have that kind of chilling effect.

Now, another area of research that has been ongoing for a long time—this is nothing new—has recently captured public attention. That is the research into cloning, cloning cells. Now, there is a man in Chicago—I don't know him and I never have met him—and his name is Richard Seed. Well, he caused quite a sensation a few weeks ago by saying he intends to clone infertile people within the next 2 years. Well, when I first heard this, I said, who is this guy? I never heard of him and I have been involved in research, medical research for a long time. Well, I found out that, quite frankly, he is a very irresponsible individual. He doesn't have the expertise himself. He doesn't have the laboratory, the money, or the wherewithal. I think most researchers and policymakers that I know who know of this person say that he is both out of the mainstream and that his plans for cloning are, at the very least, premature.

Now, again, from all that I have read—and now I have seen him on television—I think that Mr. Seed is more interested in getting his name in the paper than actually carrying out any legitimate scientific research. This is the unfortunate part of it. Why should the irresponsible actions of an individual like Mr. Seed lead to irresponsible actions on our part, because that is exactly what we are doing? Is Mr. Seed irresponsible? I believe so, absolutely. As I said, he doesn't have the expertise, the lab, or the wherewithal to even carry out this research. So he is making very irrational, irresponsible, inflammatory statements. But then why should we respond irresponsibly? I think we should respond responsibly and very carefully to an area of scientific research that can hold so much promise to alleviate pain and suffering and premature death all around the world.

Let's not act irresponsibly because one person in America has spoken irresponsibly. S. 1601, the bill we will be having a cloture vote on tomorrow, bans the use of cloning technology called somatic cell nuclear transfer. To create an unfertilized egg cell, even if this egg cell is for research, is totally unrelated to the cloning of a human being. For example, if the cell is grown under special laboratory conditions, it does not become a child, or a baby, but instead becomes specific tissue such as a muscle, nerve, or skin.

Just think of the potential of this kind of technology. I have looked into this a lot over the last several years. Science makes genetically identical tissues and organs for the treatment of a vast array of diseases.

I gave a sort of off-the-cuff set of comments last summer when this issue came up with Dolly, the sheep that was cloned in Scotland. Dr. Wilmut was at our committee. I talked about the need to continue research into cloning of cells. I said it was going to happen in my lifetime. I certainly stand here and hope that it does.

Shortly after that, I was at a restaurant in a small town in Iowa. A person came up to me, a friend of mine. I went over to their booth to see them. There was a woman there whom I had never met, a rather young woman with her husband. I was introduced to them. Just right out of the clear blue she said, "Thank you for what you said about cloning and taking the position you did on cloning." I don't even think it was in the newspaper. It was on television, I think. CNN may have carried that type of thing. But I was curious as to why this young woman, who, if I am not mistaken, lives on a farm, I believe—I can't quite remember that detail. I asked her, "Why are you so interested in this?" She said because she has a rare kidney disorder. She is hoping because of rejection possibilities that there might come a time when we could actually grow the kind of tissue that would develop into a kidney to replace her kidney so that there wouldn't be that possibility of rejection. She got it. She understood it.

That is what we are talking about. Those are the kinds of possibilities that I believe will happen in my lifetime if we do not act irresponsibly and irrationally.

This bill, S. 1601, would make it a crime to conduct some research seeking to generate stem cells to treat a wide variety of and a wide range of deadly and disabling diseases.

S. 1601 could ban blood cell therapies for diseases such as leukemia and sickle cell anemia, nerve cell therapies for Alzheimer's disease, Parkinson's disease, Lou Gehrig's disease, and multiple sclerosis. It could ban nerve cell therapy for spinal cord injuries, a very promising area of research for cloning. It could ban pancreas cells to treat diabetes, skin cell transplants for severe burns, liver cell transplants for liver damage, muscle cell therapies for muscular dystrophy and heart disease. This bill before us could ban research on cartilage cells for reconstruction of joints damaged by arthritis or injuries. It could ban cells for use of genetic therapy to treat 5,000 different genetic diseases, including cystic fibrosis, Tay-Sachs disease, schizophrenia, depression, and other diseases. S. 1601 could permanently ban all of this type of research.

In addition, under this bill, scientists could be thrown in jail for 10 years if they conduct this research—research which may not have any single thing to do with cloning a human being.

Last year, during this hearing on human cloning research, someone asked, "Are there appropriate limits to human knowledge?" Quite frankly, I responded—and I respond again—to say that I do not think there are any appropriate limits to human knowledge, none whatsoever. I think it is the very essence of our humanity and human nature. As long as science is done ethically and openly and with the informed consent of all parties, I do not think Congress should attempt to place limits on the pursuit of knowledge.

To those who suggest that cloning research is an attempt to play God, I invite you to take your ranks alongside Pope Paul V who, in 1616, persecuted the great astronomer Galileo for heresy—for saying that the Earth indeed revolved around the Sun and not otherwise.

But we don't have to go back that far. Not too long ago in our Nation's history, Americans viewed artificial insemination as abhorrent and its use was banned as being morally repugnant—even for animals; even for animals. There was an attempt to ban artificial insemination. Of course, now that is about all we use on the farm these days. Heart transplants were scorned and X-rays were considered witchcraft. But today we don't think twice about test tube babies, in vitro fertilization, or organ transplants.

Throughout the 1950s, whenever we pushed the bounds of human knowledge, there has always been a constant refrain of saying, "Stop—you are playing God." But if a couple did not have a baby and decides to seek artificial insemination, is that playing God? If a patient is dying of kidney disease and a doctor decided to transplant healthy kidneys, is that playing God? If a patient is dying of heart disease and receives a heart transplant, are we playing God?

Others say that human cloning research is demeaning to human nature. I am sorry; I don't think so. I think that any attempt to limit the pursuit of human knowledge is demeaning to human nature. I think it is the very essence of our humanity to ask how and why and if and what. I think it is demeaning to human nature to raise unfounded fears among the people of America. I think that is demeaning to human nature.

As I said, I think the finest part and the very essence of our human nature and our humanity is to ask why, how, and what if. It is our very humanity that compels us to probe the universe from the subatomic to the cosmos, and, yes, from blastocysts to the full human anatomy. Our humanity compels us to do that.

However, I must admit that I think it is rightly proper for us as policymakers to ask how human cloning research is going to affect our Nation. It is right and proper for us to examine the use of public funds for scientific research.

But I urge my colleagues to proceed with caution on this legislation. What we are talking about here is not the cloning of a human being. What we are talking about is the cloning of cells, and without further research and appropriate regulations, many people will die and become ill and spend very, very miserable lives when that could otherwise be alleviated through this cloning research.

So I have to ask: Why the rush to pass hastily drafted legislation on this very complex technical subject? We need to take the time to consider what could be the unintended consequences. The U.S. Congress and the Senate should tread very softly before sending scientists to jail for what could be promising research to cure diseases and disabilities. . . .

Mr. President, I hope that tomorrow, when we vote on this, that the Senate will choose to be on the side of the Galileos, those who want to expand human knowledge, those who will not be constricted by outmoded and outdated ideas, who understand it's the very nature of our humanity to ask how and why and what if. No, not to be on the side of those who wanted to keep the Sun moving around the Earth, but to be on the side of progress and advancement, enlightenment and unlimited human potential.

SPEECH

U.S. Senate

Senator Christopher "Kit" S. Bond

Let me be quite clear. This bill does not stop existing scientific research. I am as concerned as anyone here about the need for research on a whole range of diseases, things that can be perhaps cured or at least dealt with by stem cell research, by many other techniques that are now in progress today. Our bill does not stop any of that research. . . .

The measure places a very narrow ban on the use of somatic cell nuclear transfer to create a human embryo. That is what we are talking about. Everybody said, "We agree we shouldn't be creating a human embryo by cloning," and that is what this bill does.

Over the past week, we have had a lot of distortion and, unfortunately, inflamed rhetoric by some of the big special interests, the likes of which I have not seen in my many years of public service. We have asked our opponents on numerous occasions, we have sat down with them, Senator Frist, Senator Gregg, our staffs and I sat down and said, "OK, if we all agree we shouldn't be creating a human embryo by cloning, how do you want to tighten it up?"

They are not willing to come forward because there are some rogue scientists, maybe some big drug companies, big biotech companies, who want to create human embryos by cloning. They think that would be a great way to be more profitable, to do some research on cloned human embryos. I think that is where we need to draw the line.

Senator Christopher "Kit" S. Bond (R) is a member of the U.S. Senate from Missouri. He served two terms as Missouri governor before winning election to the Senate in 1986, where he was reelected in 1992 and 1998. He was the leading sponsor of the bill to ban human cloning in 1998. Selections from *Congressional Record*, February 11, 1998.

People say we want to have hearings. We have had hearings on the whole issue last year. We have debated it, and it comes down to the simple point: Do you want to say no to creating human embryos by cloning, by somatic cell nuclear transfer, or do you want to say, as my colleague from California would in her bill, "Oh, it's fine to create those human embryos by somatic cell nuclear transfer, so long as you destroy them, so long as you kill those test tube babies before they are implanted"?

There are a couple problems, very practical problems. Once you start creating those cloned human embryos, it is a very simple procedure to implant them. Implantation of embryos is going along in fertility research now, and it would be impossible to police, to make sure they didn't start implanting them.

But even if the objectives of the bill of my California colleague were carried out, it would mean that you would be creating human embryos by cloning, researching with them, working with them and destroying them. Do we want to step over that ethical line? I say no.

It is not going to be any clearer 3 months from now, 6 months from now than it is now. What is going to be different is that in 3 or 6 months, the rogue scientist in Chicago or others may well start the process of cloning human embryos by somatic cell nuclear transfer. That is why we say it is important to move forward on this bill.

If we bring this bill to the floor, we are happy to listen to and ask for specific suggestions from those who are concerned about legitimate research, but we have been advised time and time again that there is no legitimate research being done now in the biotech industry that uses somatic cell nuclear transfer to clone and create a human embryo as part of the research on any of these diseases.

We have heard from patient groups, people who are very much concerned, as we all are, about cancer, about juvenile diabetes, cystic fibrosis, Alzheimer's—the whole range of diseases. We can deal with those diseases. We can deal with the research without cloning a human embryo.

The approach of my colleagues from California and Massachusetts would lead us down the slippery slope that would allow the creation of masses of human embryos as if they were assembly line products, not human life. How would the Federal Government police the implantation of these human embryos?

By allowing the creation of cloned test tube babies so long as they are not implanted, our opponents' bill calls for the creation, manipulation and destruction of human embryos for research purposes.

I have a letter that I will enter into the RECORD from Professor Joel Brind, Professor of Human Biology and Endocrinology at Baruch College, The City University of New York. . . . I quote from a portion of it:

> Industry opponents also correctly point out that S. 1601 would ban the production of human embryos for research or other purposes entirely unrelated to the aim of cloning a human being. And well it should. . . . In fact, it is in this area of research and treatment, to wit, the generation of stem cells, from which replacement tissues or organs could be produced for transplantation into the patient from whom the somatic cell originally came, which is most important to the biotech industry, for obvious reasons. For reasons just as obvious to anyone with any moral sense, such practices must be outlawed, for otherwise, our society would permit the generation of human beings purely for the purpose of producing spare parts for others, and thence to be destroyed. Some may call this a "slippery slope"—I believe "sheer cliff" would be more accurate.

LIBERALISM AND CLONING

Adam Wolfson

The summer of 1998 will be remembered not only for baseball sluggers and Oval Office sex but also for animal cloning. Reports of cloned cows and mice and even the cloning of a nearly extinct breed of New Zealand cow trickled in from around the globe. Curiously, though, the new spate of clonings brought a reaction different from the concern that greeted the announcement almost two years ago of a sheep named Dolly—the first animal ever cloned from an adult mammal.

The general response to Dolly was, How do we stop this thing from being done to humans? President Clinton promptly banned federal funding of human cloning and asked the private sector to go along voluntarily. Warning scientists against playing God, the president ordered the National Bioethics Advisory Commission to investigate the question and report in 90 days. The commission's report called the cloning of humans "morally unacceptable for anyone in the public or private sector"—though it recommended only a temporary ban, pending scientific advances and public debate. (Skeptics wondered whether the commission were only buying time until scientists could perfect a cloning technique, making it safe to try on humans in due course.)

In June 1997, the president sent Congress legislation that would prohibit the cloning of humans for five years. And there was every reason to think that such a bill would pass. The vast majority of Americans oppose human cloning, and 20 European states have banned it. But little happened, at least until, in December 1997, the scientist Richard Seed announced his plans to clone a human being. The FDA quickly asserted its regulatory authority to

This essay originally appeared in the *Weekly Standard* on October 5, 1998.

stop anyone who would attempt to clone a human, and there was a flurry of activity on Capitol Hill. But again, nothing happened.

Conservative Republicans were ready to act, but they wanted to outlaw both human cloning and experiments on human embryos. Democrats were willing to back the former goal, but they (and some Republicans) argued that a ban on embryo research would block advances against cancer, Parkinson's, Alzheimer's, diabetes, and other diseases. The upshot: no legislative ban.

By this summer, the concern had somehow dissipated. The public seemed blasé. In August, a wealthy Texan reportedly donated $2.3 million to researchers at Texas A&M University to clone his pet dog, Missy. And Dr. Lee M. Silver, a Princeton molecular biologist and the author of *Remaking Eden: Cloning and Beyond in a Brave New World,* told the *Washington Post,* "In six years, you'll be calling me to ask me what I think about the first human clone."

Why did nothing happen? How did we move so casually from calls for a five-year moratorium on human cloning to predictions that a human will be cloned within roughly the same time? Part of the answer lies in the transformation of liberalism over the last quarter century from a public philosophy with a vibrant moral center and real intellectual ballast to one that is merely the shell of its former self.

The debate about human cloning made its first appearance in the 1970s. James D. Watson and Francis Crick had published their work on the structure of DNA in 1953, and scientists had had some limited success cloning frogs in the 1950s and 1960s. By the early 1970s, the wisdom of human cloning began to be widely discussed, by Watson in the *Atlantic Monthly* (against), Nobel laureate Joshua Lederberg in the *Washington Post* (for), and Leon Kass in the *Public Interest* (against). In the debate that ensued, some of liberalism's most important spokesmen raised their voices against human cloning, including the *New York Times Magazine* and liberalism's leading constitutional scholar, Laurence H. Tribe.

In 1972, the *New York Times Magazine* published a powerful critique of human cloning by Columbia University psychiatrist Willard Gaylin. In his article, Gaylin gave expression to the principal *liberal* reasons for opposition. He articulated a principle of "humanness." Science offers mankind endless possibilities, he argued, but we do not choose to travel down those paths that we deem inconsistent with our given nature. So too Tribe, in a series of articles in the early '70s, opposed cloning on the grounds that it would detract from our "humanness." It would do so in several ways.

As many commentators have pointed out, cloning is but a step along the way towards the genetic manipulation of the species. Dr. Silver argues that it is cloning that will take genetic engineering out of the realm of fantasy and make it a reality. But man's ability to reconstruct human nature is, Gaylin pointed out, "by definition, the capacity to destroy himself through transformation into another creature—perhaps better, but not man."

In the choice between man and superman, Gaylin affirmed liberalism's faith in man. And I would point out that whether the eugenics is organized from above, by, say, a totalitarian state, or opted for from below, by yuppies who wish to improve, by genetic manipulation, the SAT scores or physical appearance of their offspring, the practice is objectionable. (Obviously, we are not talking here about uncontroversial efforts to ensure health.) Liberalism's opposition to the eugenic transformation of man into superman was always based on more than whether the process was state directed. The objection was to the thing itself—to the very idea of creating a superman.

Other arguments against cloning also flowed from this principle of humanness. For instance, Tribe in the early 1970s emphasized that cloning and similar technologies would turn human beings into mere objects. Certainly, many technologies are supportive of human dignity, but, as Tribe argued, human cloning and related technologies are not among them. By making human nature increasingly subject to external, scientific control, such technologies would make it difficult to conceive of the resulting products as "free and rational" agents. The upshot, as Tribe then saw, would be disastrous: "In a society that came to view its members as just so many cells or molecules to be manufactured or rearranged at will, one wonders how easy it would be to recall what all the shouting about 'human rights' was supposed to mean."

Another anti-cloning principle articulated by Gaylin had to do with liberalism's core belief in the value of life and the right to life. Any attempt to clone human beings, he warned, would produce failures along the way. And what, he asked,

> will we do with the discarded messes along the line? What will we do with those pieces and parts, near-successes and almost-persons? What will we call the debris? At what arbitrary point will the damaged "goods" become damaged "children," requiring nurture rather than disposal? The more successful one became at this kind of experimentation, the more horrifyingly close to human would be the failures. The whole thing seems beyond contemplation for ethical and esthetic, as well as scientific, reasons.

This danger strikes at liberalism's heart. At the point that liberalism begins to make distinctions between human life that is worthy of preservation and that which is not, entitling some to rights but not others, it ceases to be entirely liberal.

Finally, Gaylin and Tribe made the connection between liberalism's concern to preserve the natural environment and what should be its concern to preserve man's nature. On this point, one would do well to recall that Gaylin's and Tribe's articles were written soon after the environmental movement took off. In 1962, Rachel Carson's *Silent Spring* was published, and in 1970, Sen. Gaylord Nelson kicked off Earth Day. The parallel between preserving the planet and safeguarding man was obvious to Gaylin: "The unpredicted complexities of environmental intervention, with the resulting ecological disasters, should serve as a warning model." What Rachel Carson had argued DDT was to the natural environment, Gaylin suggested cloning was to human nature. What was the use of protecting Mother Nature if one did nothing to preserve human nature?

That was liberalism circa 1972. Twenty-five years later, liberalism has all but forgotten its powerfully reasoned arguments against human cloning. After the February 1997 announcement of Dolly's birth, Gustav Niebuhr began a front-page news story for the *New York Times* as follows: "The cloning of an adult mammal offers a striking example of how technology can outpace the moral and social thinking that would guide it." The *Times*'s lead editorial just a few days before had begun with the same observation: "The startling news that scientists have cloned an adult sheep . . . is a reminder that reproductive technologies are advancing far faster than our understanding of their ethical and social implications." Had the writers at the *Times* failed to search their own archives? Or was it simply that they had rejected liberalism's old understanding of the ethics of human cloning and gone searching for new understandings?

In the years between Gaylin's prophetic article and Dolly's birth, liberalism underwent a transformation. To begin with, liberalism came to advocate an absolute right to abort a fetus (even if exercising this right meant performing the ghastly procedure known as partial-birth abortion on a third-trimester fetus). For this reason alone, liberalism would eventually have to say yes to the research on human embryos that will make cloning possible: On what principled ground could one insist on the right to abort a fetus while objecting to experiments on it?

Gaylin himself predicted that liberalism's embrace of abortion would have "broad social effects" on issues of life and death. And he was right. In the early 1970s, before *Roe v. Wade* had wrought its effects, liberals could not help but shudder when confronted by Gaylin's chilling question, What is to be done with cloning's "discarded messes," its not-quite-human failures? Not so today. Liberalism has come to view such "discarded messes" as a means for conquering disease. The reason liberals in Congress gave for opposing the Republican-proposed ban on cloning, including the cloning of embryos, was that it would slow or even halt medicine's advance on many important fronts. One must admire the humanitarian impulse behind such scientific endeavors. Who can lightly turn his back on cures for terrible diseases? And as Gaylin acknowledged, it is very difficult to say what precisely is lost in the bargain.

But one can wonder, as Gaylin did, whether an unrestrained effort to relieve suffering will eventually vitiate our very humanity. Already there are signs of such an effect. In the matter of doctor-assisted suicide, for example, well-meaning liberals advocate the killing of the sick and dying as a way of ending undeniable anguish. An unrestrained fear of death and suffering apparently makes us fearless and heartless. Even the idea of humanness itself, on which Gaylin and Tribe built their case against cloning, has lost its moral content. In the summer of 1997, a group calling itself the International Academy of Humanism issued a declaration defending human cloning and related research for its "potential benefits." These newfangled "humanists," including the late Isaiah Berlin, Harvard philosopher W. V. Quine, and Harvard sociobiologist Edward O. Wilson, found the question of human cloning no more morally "profound" than that of computer encryption.

But if Gaylin foresaw true humanism's many vulnerabilities, he could not have imagined what happens when liberals go postmodern. Such liberals are, at best, indifferent to the question of human cloning; at worst, they view cloning technology as something to be embraced in the name of liberation and experimentation. Today's cloning technology thus heralds a new ideological force in American politics. Once suspicious of technology—think of liberal opposition to nuclear power—liberals of a postmodern persuasion will ally themselves with the new technologies of human cloning and genetic engineering because of the liberating potential of those technologies. And thus will experiments in science come to the aid of lifestyle experiments.

For example, in a remarkable piece published in the *New York Times* in December 1997, and in an elaboration published in *Clones and Clones: Facts and Fantasies about Human Cloning,* Laurence Tribe turned his back on his old liberal arguments against cloning. "Who was I—who is anyone—to forecast *which* technologies will over time generate transformations sufficiently deep . . . that we may confidently favor the[ir] outright prohibition?" he now exclaimed. "How can any of us feel so confident that the meaning of humanity will be degraded by human cloning in any and all circumstances that we are prepared to shut [it] down?"

Having gone postmodern, Tribe now finds it impossible to distinguish between what degrades man and what supports his dignity. Instead of worrying about treating humans as objects, or destroying human nature, as he did 25 years ago, he now rails against "essentialism" and puts the words "natural," "unnatural," and "human nature" in ironic quotation marks. And, chastising the opponents of human cloning, he warns that "when fear of the unnatural drives a campaign to ban some innovation, then that very fear may be more fearsome than the innovation that spawns it."

One of the great divides between old liberals and their postmodern cousins is just this lack of belief in human nature. To postmodernists, the idea of human nature is nothing more than an outdated social construct used to oppress "difference." As the postmodernist Richard Rorty once wrote, "There is nothing deep inside each of us, no common human nature. . . . There is nothing to people except what has been socialized into them." Thus the importance to postmodernists of personal experimentation and "self-creation."

Yet, what is harmless wordplay in the mouths of college philosophy professors, perhaps at most encouraging impressionable young men and women to do things they otherwise might not have done, becomes in Tribe's hands a reason for embracing human cloning. As he puts it: "A society that bans acts of human creation that reflect unconventional sex roles or parenting models . . . for no better reason than that such acts dare to defy 'nature' and tradition . . . is a society that risks cutting itself off from vital experimentation and risks sterilizing a significant part of its capacity to grow." Why of course! If we are for "self-creation" and "vital experimentation," what could be more wonderfully "transgressive," as the postmodernists like to say, than the technology of human cloning and eventually of genetic engineering?

Now that the old liberal arguments against human cloning have fallen into desuetude, it's not clear what kind of political suasion can stop it from eventually happening. Of the three main elements of the conservative coalition—social or religious conservatives, libertarians, and the business class—only religious conservatives have expressed any serious opposition to cloning. The business class sees in the burgeoning cloning industry money to be made. As *New York Times* reporter Gina Kolata points out in her book *Clone: The Road to Dolly and the Path Ahead*, we owe little thanks (if that is what is owed) to government or university scientists for the technology of cloning. Most of the scientists who gave us animal cloning worked in industry; and their aim was not to save man or uncover new knowledge but to make a buck.

As for libertarians, they make the same old arguments against a moratorium on cloning that they have long made against the government's passing laws against drugs or pornography or gambling: It's none of the state's business. Some of them have spiced this old libertarian argument with a dash of techno-enthusiasm. They see in cloning new possibilities for the improvement of "man." Or they argue that any sort of ban on human cloning would be pointless since law and politics can never stop scientific and technological progress—and they may be right. But what's the consolation in that, since a science that inevitably overcomes law and politics can also overcome the political liberties that libertarians, liberals, and conservatives all support?

However that may be, our intellectual disarmament before the science of cloning leaves human cloning likely to occur in the near future. Since the liberal arguments against human cloning no longer resonate, perhaps a more ancient treatment of the subject is in order. In *Paradise Lost*, Milton describes how "Sin" springs forth from Satan's head; whereupon Satan falls in love with the creature—who is both his "perfect image" (his clone?) and his daughter. Then Satan copulates with "Sin," impregnating her with "Death"—whom he will later call "thou Son and Grandchild both." If the cloners have their way, we too shall someday speak such frightening words, being both parents and siblings to our children.

TESTIMONY

U.S. Senate

James A. Thomson

My name is James Thomson and I'm a developmental biologist at the University of Wisconsin-Madison. My group has recently reported the derivation of human Embryonic Stem (ES) cells that can proliferate indefinitely in tissue culture and yet maintain the potential to form many, and possibly all, adult cell types. Human ES cells thus provide a potentially unlimited source of specific differentiated cell types for basic biological research, pharmaceutical development, and transplantation therapies. These human ES cell lines were derived from in vitro fertilized embryos before the formation of any fetal structures. These embryos were produced by in vitro fertilization for clinical purposes, but were in excess of clinical needs and were donated after informed consent. The informed consent process was detailed and specified the purpose of the research, its context, and its implications.

Human ES cell lines are important because they could provide large, purified populations of human cells such as heart muscle cells, pancreatic cells, or neurons for transplantation therapies. Many diseases, such as juvenile onset diabetes mellitus and Parkinson's disease, result from the death or dysfunction of just one or a few cell types, and the replacement of those cells by transplantation could offer lifelong treatment. Human ES cells are also important because they will offer insights into developmental events that cannot be studied directly in the intact human embryo, but which have important

James A. Thomson is a developmental biologist and professor of anatomy at the University of Wisconsin–Madison. He directed the group that first isolated embryonic stem cells in 1998 and was a finalist for the World Technology Award in 1999. Selections from testimony before the U.S. Senate Committee on Appropriations, Subcommittee on Labor, Health, Human Services, and Education, December 2, 1998.

consequences in clinical areas, including birth defects, infertility, and pregnancy loss. Screening tests that use specific ES cell derivatives will allow the identification of new drugs, the identification of genes that could be used for tissue regeneration therapies, and the identification of toxic compounds.

Although the long-term potential for human therapies resulting from human ES cell line research is enormous, these therapies will take years to develop. Significant advances in developmental biology and transplantation medicine are required, but I believe that therapies resulting from human ES cell research will become available within my lifetime. How soon such therapies will be developed will depend on whether there is public support of research in this area. Private companies will have an important role in bringing new ES cell-related therapeutics to the marketplace; however, the current ban in the U.S. on the use of Federal funding for human embryo research discourages the majority of the best U.S. researchers from advancing this promising area of medical research. . . .

WHAT ARE HUMAN EMBRYONIC STEM CELLS?

In the adult mammal, cells with a high turnover rate are replaced in a highly regulated process of proliferation, differentiation, and programmed cell death from undifferentiated adult "stem cells." Tissues from which stem cells have been extensively studied include blood, skin, and the intestine. In the human small intestine for example, approximately one hundred billion cells are shed and must be replaced daily. Although various definitions have been proposed, characteristics of adult stem cells generally include: (i) prolonged proliferation, (ii) self-maintenance, (iii) generation of large numbers of progeny with the principle phenotypes of the tissue, (iv) maintenance of developmental potential over time, and (v) the generation of new cells in response to injury. Thus, stem cells in the adult sustain a relatively constant number of cells and cell types. Several properties of adult stem cells limit their therapeutic potential. First, all adult stem cells are committed to becoming a relatively restricted number of cell types. Second, although adult stem cells can divide for prolonged periods, cell division can occur only a finite number of times, so there is a limit to how much they can be expanded in tissue culture. Third, the sustainable culture of adult stem cells has not yet been achieved. And fourth, several tissues of clinical importance, such as the heart, completely lack stem cells in the adult. Thus, after a heart attack, when heart muscle dies, there is no regeneration of heart muscle, only the formation of non-functional scar tissue.

In contrast to the adult, embryonic cell proliferation and differentiation elaborates an increasing number of cells and cell types. In mammals, each cell of the cleavage stage embryo has the developmental potential to contribute to any embryonic or extraembryonic cell type, but by the blastocyst stage, cells on the outside of the embryo (the trophectoderm) are committed to a particular cell type found in the placenta. The cells on the inside of the blastocyst (the inner cell mass) contribute to all the tissues of the embryo proper. Soon after the blastocyst stage, cells of the inner cell mass develop into cells that are developmentally restricted to particular lineages. Because the cells of the inner cell mass proliferate and replace themselves in the intact embryo for a very limited time before they become committed to specific lineages, they do not satisfy the criteria for stem cells that are applied to adult tissues. In contrast, if the inner cell mass is removed from the normal embryonic environment and dissociated under appropriate conditions, the cells will remain undifferentiated, replace themselves indefinitely, and maintain the developmental potential to contribute to all adult cell types. Thus, these inner cell mass-derived cells satisfy the criteria for stem cells outlined above, and they are referred to as embryonic stem (ES) cells. The derivation of mouse ES cells was first reported in 1981. We have recently described the isolation of human ES cell lines that satisfy the following criteria for ES cells: (i) derivation from the pre- or per-implantation embryo, (ii) prolonged undifferentiated proliferation, and (iii) stable developmental potential after prolonged culture to form differentiated derivatives of all three embryonic germ layers (endoderm, mesoderm, and ectoderm) which are the three basic lineages that give rise to all of the cells of the adult. . . .

WHY ARE HUMAN ES CELLS IMPORTANT?

Human ES cell lines could offer insights into developmental events that cannot be studied directly in the intact human embryo or in other species, but which have important consequences in clinical areas, including birth defects, infertility, and pregnancy loss. Particularly in the early post-implantation period, knowledge of normal human development is largely restricted to the description of a limited number of sectioned embryos and to analogies drawn from the experimental embryology of other species. Although the mouse is the mainstay of experimental mammalian embryology, early structures including the placenta, extraembryonic membranes, and the egg cylinder all differ significantly from those of the human embryo. Human ES cell lines will be particularly valuable for the study of the development and function of tissues that differ between mice and humans.

Elucidating the mechanisms that control differentiation will facilitate the directed differentiation of ES cells to specific cell types. The standardized production of large, purified populations of normal human cells such as heart muscle cells and neurons will provide a potentially limitless source of cells for drug discovery and transplantation therapies. For example, large purified populations of ES cell-derived heart muscle cells could be used to find new drugs to treat heart disease. Many diseases, such as Parkinson's disease and juvenile onset diabetes mellitus, result from the death or dysfunction of just one or a few cell types. The replacement of those cells could offer lifelong treatment. In addition to Parkinson's disease and juvenile onset diabetes, the list of diseases potentially treated by this approach is long, and includes myocardial infarction (heart muscle cells and blood vessels), atherosclerosis (blood vessels), leukemia (bone marrow), stroke (neurons), burns (skin) and osteoarthritis (cartilage). For the foreseeable future, these therapies will involve the repair of organs by the transplantation of cells or simple tissues, but not the replacement of entire organs.

Strategies to prevent immune rejection of the transplanted cells need to be developed, but could include banking ES cell lines with defined major histocompatibility complex backgrounds or genetically manipulating ES cells to reduce or actively combat immune rejection. Significant advances in basic developmental biology are required to direct ES cells efficiently to lineages of human clinical importance. However, progress has already been made in the in vitro differentiation of mouse ES cells to neurons, hematopoietic cells, and cardiac muscle.

WHY IS THE DERIVATION OF HUMAN ES CELL LINES CONTROVERSIAL?

The derivation of human ES cells lines is controversial both because of the use of human embryos, and because of the properties of ES cell lines. . . . The use of human embryos generated by in vitro fertilization for any research purpose raises complex ethical issues, and it is beyond the scope of this testimony to review the wide range of views on this subject. Research on human preimplantation embryos has been reviewed in depth by national panels in Britain, Canada, and the United States. In Britain, some research on human preimplantation embryos is allowed, but it is very carefully monitored and regulated, regardless of whether public or private funds are used. The derivation of human ES cell lines is already being publicly funded in Britain. In the U.S., the most recent and complete review of human preimplantation embryo research was completed by the National Institutes of Health (NIH) Human Embryo Research Panel in the fall of 1994. Although the guidelines suggested by the NIH panel do not have the force of law, we followed those guidelines in our derivation of human ES cell lines, as no other Federal guidelines currently exist for privately funded human embryo research. The derivation of human ES cell lines was specifically addressed by the NIH Panel, and it was recommended as acceptable for Federal funding, as long as embryos were not fertilized expressly for that purpose. The subsequent ban on Federal funding for research on human preimplantation embryos has prevented the NIH from funding the derivation of human ES cell lines, and has prevented the NIH from regulating and monitoring research in this area. The NIH Panel both recognized the therapeutic promise of human embryo research and recognized that the human embryo warrants serious moral consideration as a developing form of human life. For these reasons, the panel concluded in part, "It is in the public interest that the availability of Federal funding and regulation should provide consistent ethical and scientific review for this area of research. The Panel believes that because the preimplantation embryo possesses qualities requiring moral respect, research involving the ex utero preimplantation human embryo must be carefully regulated and consistently monitored."

Human ES cell lines are not the equivalent of an intact human embryo. If a clump of ES cells was transferred to a woman's uterus, the ES cells would not implant and would not form a viable fetus. The recent cloning of sheep and mice by the transfer of an adult cell nucleus to an enucleated oocyte demonstrates that at least some adult cells are totipotent (capable of forming an intact embryo that is capable of developing to term). If nuclear transfer from adult cells to enucleated oocytes allows development to term in humans, then the transfer [of] a nucleus from an ES cell to an enucleated oocyte might also result in a viable embryo. However, if someone wanted to clone a specific famous or infamous individual, the transfer of a nucleus from [an] adult cell would have to be used, not a nucleus from an ES cell. One of the major uses of mouse ES cells has been to genetically modify the germ line (gametes) of mice in very specific ways. However, because of the significant reproductive differences between mice and humans, and the inefficiency of this method of modifying the germ line, human ES cells do not increase the potential for modifying the human germ line. As has already been demonstrated in domestic animal species such as the cow and sheep, nuclear transfer techniques allow a much more efficient way to modify the germ line of species with long generation times. Thus, if someone wanted to genetically modify the human germ line, there are already other approaches that would be quicker and more efficient than using human ES cells. The NIH Human Embryo Research Panel included recommendations against research involving nuclear transfer to enucleated human oocytes or zygotes followed by transfer to a uterus, and any formation of chimeras with human embryos; these recommendations would effectively preclude the use of ES cells for human cloning or modifying the human germ line.

TESTIMONY

U.S. Senate

Richard M. Doerflinger

In discussions of human experimentation, the researcher's temptation is to think that if something technically can be done it ethically should be done—particularly if it may lead to medical benefits or advances in scientific knowledge. A civilized society will appreciate the possibilities opened up by research, but will insist that scientific progress must not come at the expense of human dignity. When this important balance is not maintained, abuses such as the Tuskegee syphilis study or the Cold War radiation experiments become a reality.

In deciding whether to subsidize various forms of human experimentation, legislators are not merely making an economic decision to allocate limited funds. On behalf of all citizens who pay taxes, they are making a moral decision. They are declaring that certain kinds of research are sufficiently valuable and ethically upright to be conducted in the name of all Americans—and that other kinds are not. By such funding decisions, government can make an important moral statement, set an example for private research, and help direct research toward avenues which fully respect human life and dignity as they seek to help humanity.

Three kinds of experiments involving human embryos or embryonic cells have recently come to public attention. On one level, some of these experiments advance the ethical and legal debate on human cloning. They indicate

Richard M. Doerflinger is associate director for policy development at the Secretariat for Pro-Life Activities, United States Conference of Catholic Bishops. He has testified on human embryo research before Congress, the National Institutes of Health, and the National Bioethics Advisory Commission. Selections from testimony before the U.S. Senate Committee on Appropriations, Subcommittee on Labor, Health, Human Services, and Education, December 2, 1998.

that cloning is not necessary for promising stem cell research, and thus that it may be banned without endangering such research. At the same time, however, each of these experiments raises ethical problems of its own.

Currently, the drive for advances in human fetal and embryonic research is balanced against ethical considerations in three significant areas of federal law. It is important to review these to address the question: What moral principles are reflected in these enactments that can help us to make a moral judgment on new experiments that may not have been anticipated before?

1. *Live fetal research is governed by federal regulations on the protection of human subjects first issued in 1975.* Federal regulations on fetal research treat the prenatal human being as a human subject worthy of protection, from the time of implantation in the womb (about one week after fertilization) until a child emerges from the womb and is found to be viable. Essentially the same standard is applied here as in regulations protecting live-born children: Since the unborn child is a helpless subject incapable of giving informed consent to experimentation, federally funded research involving this child is permissible only if (a) it could be therapeutic for that particular child (as with prenatal surgery to correct congenital defects), or (b) it is necessary to obtain important information and will not subject the child to significant risk of harm. In 1985 Congress further clarified this standard through an amendment to the National Institutes of Health reauthorization act: In assessing research on live fetuses in utero, protection from risk must be "the same for fetuses which are intended to be aborted and fetuses which are intended to be carried to term." No matter what fate may be planned for the developing human being by others, the government must still make its own moral decision to respect life—it cannot single out certain lives as disposable, or as uniquely fit for harmful research, simply because someone else plans to show disrespect for those lives.

2. *Embryo research involving human embryos outside the womb—such as embryos produced in the laboratory by in vitro fertilization (IVF) or cloning—has never received federal funding.* Originally this was because the federal regulations of 1975 prevented funding of IVF experiments unless such experiments were deemed acceptable by an Ethics Advisory Board—and after the first such board produced inconclusive results in 1979, no Administration chose to appoint a new board. In 1994, after this regulation was rescinded by Congress, a Human Embryo Research Panel recommended to the National Institutes of Health that certain kinds of harmful nontherapeutic experiments on human embryos receive federal funding—but the Panel's recommendations were rejected in part by President Clinton, then rejected

in their entirety by Congress. Since 1995, three successive Labor/HHS appropriations bills have prevented federal funding of experiments which involve (a) creating human embryos for research purposes, or (b) subjecting human embryos in the laboratory to risk of harm or death not permitted for fetuses in utero under the regulations on protecting human subjects. Since 1997 this rider has explicitly banned funding of experiments involving embryos produced by cloning using human body cells.

3. Fetal tissue transplantation research has been a matter of extended controversy. Such research could receive federal funds during the Bush Administration only if the tissue was obtained from sources other than induced abortion. The possible use of ovaries from aborted fetuses to create research embryos provoked more controversy within the NIH Human Embryo Research Panel than perhaps any other proposal; in the end the Panel decided to defer any possible funding of such research until further discussion could take place. Under current federal funding policy, human fetal tissue—defined as "tissue or cells obtained from a dead human embryo or fetus after a spontaneous or induced abortion, or after a stillbirth,"—may be used for "therapeutic purposes" only if various safeguards are followed to ensure that the researcher avoids participating in an abortion and has no effect on the "timing, method, or procedures used to terminate the pregnancy."

In our view, current safeguards on the use of fetal tissue are inadequate. The only sure way to prevent federally funded research from collaborating in and providing legitimacy for abortion is to forbid abortion as a source for potentially "therapeutic" tissue. Certainly, it would be wrong for Congress to apply to early human embryos any policy less protective than that now applied to the later embryo and fetus. Existing law explicitly applies to human embryos, and destroying or discarding an embryo in the laboratory is the moral equivalent of abortion. As Congress has already done in the case of live fetal research, it should make clear in this area of research that the same standards apply to human embryos whether inside or outside the womb.

Current law on live fetal and embryonic research is no mere political compromise. It is a reflection of universally accepted ethical principles governing experiments on human subjects—principles reflected as well in the Nuremberg Code, the World Medical Association's Declaration of Helsinki and other statements. Members of the human species who cannot give informed consent for research should not be the subjects of an experiment unless they personally may benefit from it, or the experiment carries no significant risk of harming them. Only by such ethical principles do we prevent treating people as things—as mere means to obtaining knowledge or benefits for others.

Some will be surprised that such protections can exist under the U.S. Supreme Court's abortion decisions. But the Court has never said that government may not protect prenatal life outside the abortion context. It has even allowed states to declare that human life begins at conception, and that it deserves legal protection from that point onward—so long as this principle is not used to place an undue burden on a woman's "right" to choose abortion before viability. Although states may not place meaningful restrictions or prohibitions on abortion under current Supreme Court jurisprudence, harmful experiments on human embryos are illegal in ten states regardless of how they are funded. Public sentiment also seems even more opposed to public funding of such experiments than to funding of abortion.†

Moreover, a scientific consensus now recognizes the status of the early human embryo, and the continuity of human development from the one-celled stage onward, to a greater extent than was true even a few years ago. In the 1970s and 1980s, some embryologists spoke of the human embryo in its first week or two of development as a "pre-embryo" and claimed it deserved less respect than embryos of later stages. But most embryology textbooks have now dropped the term, and some texts openly refer to it as a "discarded" and "inaccurate" term.†† The Human Embryo Research Panel and the National Bioethics Advisory Commission have both rejected the term; they describe the human embryo, including the one-celled zygote, as a living organism and "a developing form of human life."

How is this human life treated in each of the three most recent developments in human embryo research?

UNIVERSITY OF WISCONSIN: STEM CELLS FROM AN IVF EMBRYO

The University of Wisconsin proposal seems to be exactly the kind of experiment that the federal funding ban was consciously directed against. Researchers obtained 36 live human embryos from IVF clinics, grew them to the blastocyst stage, and then destroyed them for their stem cells; cells from

†A national Tarrance poll in 1995 showed 18 percent support for using tax dollars for experiments that would involve destroying or discarding live human embryos in the first two weeks of development. Seventy-four percent of the Americans in the survey opposed such funding, with 64 percent strongly opposed.

††Professor Lee Silver of Princeton University, a proponent of cloning and embryo research, recently declared that the term "pre-embryo" was embraced by IVF researchers "for reasons that are political, not scientific" in an effort to "allay moral concerns" about their research.

14 of the embryos were placed in culture, and "cell clusters" from five were successfully cultured to grow tissue. The researchers report that the inner cells were "isolated by immunosurgery" from the rest of the embryo. The effect is the same as if one were to "isolate" the heart and lungs from an adult human—the being from whom the cells are taken is killed.

This kind of experiment was recommended for federal funding in 1994 by the Human Embryo Research Panel, but rejected by Congress every year from 1995 to the present. In this respect it does not present a new issue, for Congress has already decided that even so-called "spare" embryos from IVF clinics should not be subjected to destructive experiments using federal funds.

Two new issues have been raised regarding this experiment, however.

First, could the embryonic cells that are removed from these human embryos, once isolated, be seen as human embryos themselves? The question arises because these inner cells are often described as the cells that would ultimately form the "embryo proper" as development continues. If removed from the original embryo but transferred to the nurturing environment of the womb, would each cell or each cluster of cells begin to develop as a new organism? Is a special environment provided by researchers to suppress such development and divert it toward undifferentiated growth as tissue instead? Certainly such diversion of embryonic development by use of molecular signals has been proposed by some researchers. If that were at work here—if the experiment creates new embryos and then suppresses their development—funding such an experiment might also violate the current ban on creating human embryos for research purposes.

Second, what of the prospect of funding research that would use this tissue for supposedly therapeutic purposes after it has been grown in culture? Here, the ethical principles reflected in current law on fetal tissue argue against funding the research. One must refer here to the principles rather than to the exact letter of the law because, while it applies to embryos as well as fetuses, it speaks of induced abortion rather than of destroying embryos by dissection. But there seems to be no reason why the same ethical standard should not apply. Human embryos are destroyed precisely to obtain this tissue, and the timing and manner of the destruction are tailored to obtaining this kind of tissue. An effective separation between the destructive act and the harvesting of the tissue, which federal law requires in the case of tissue from an induced abortion, does not seem to exist here.

One positive development, however, is that this line of research has put to rest the claim made last year by some biotechnology companies that production of human embryos by cloning (somatic cell nuclear transfer) is necessary to develop therapies based on embryonic stem cells. Such claims assumed that adults could not be treated with such cells unless the embryos were produced by cloning to create a genetic "match" and avoid tissue rejection. But in his commentary on the Wisconsin experiment, John Gearhart has cited three other avenues, some of which were also cited by the National Bioethics Advisory Commission last year: Stem cells can be banked from multiple cell lines to prevent such reactions; they can be genetically altered to produce a universal donor line; or they can be customized using the relevant histocompatibility genes from the intended recipient.

Whatever else may be said of this research, then, it means that proposed federal bans on human cloning need no longer be held hostage to the debate on stem cell research. But the experiment itself is unethical and should not be funded. Instead, as the National Bioethics Advisory Commission has already observed, avenues should be explored for creating stem cell lines without creating or destroying human embryos.

JOHNS HOPKINS: STEM CELLS BASED ON PRIMORDIAL GERM CELLS FROM INDUCED ABORTION

Presumably the Johns Hopkins University study could not be funded unless it follows the provisions of current law regarding fetal tissue from induced abortions. We wish to reiterate here that we find the existing policy inadequate and would support federal funding only if the cells can be obtained from sources other than induced abortion.

The new question raised here is this: Are the primordial germ cells obtained from abortion victims being used to create human embryos, which are then destroyed or suppressed to provide tissue? Even the NIH Human Embryo Research Panel did not recommend funding such an experiment, and it would clearly be forbidden by the current embryo research ban.

There is some ambiguity in current reports of the new research, because the researchers speak of collecting "embryoid bodies" from these cultures and finding "derivatives of all three embryonic germ layers" in the culture. They add that some of these bodies form "complex structures closely resembling an embryo during early development," and that they "appear to recapitulate the normal developmental processes of early embryonic stages and promote the cell-cell interaction required for cell differentiation."

However, if this research is now conducted—or could be conducted—to establish useful cell lines without creating early human embryos, it would avoid some of the serious ethical problems associated with other experiments in this field. In that case the only remaining ethical problem is the use of cells from induced abortion, which does not seem necessary to the nature of the research. We urge that the use of cells from spontaneous abortions, ectopic pregnancies or other sources be explored instead.

ADVANCED CELL TECHNOLOGY: HUMAN CLONING USING COWS' EGGS

While this third type of experiment has not been fully reported in the medical literature it seems to pose a relatively new question—that of human/animal hybrids—as well as an old one, that of somatic cell nuclear transfer (cloning) to make human embryos for research purposes. Even the Human Embryo Research Panel opposed funding the former; the current ban on embryo research rightly forbids funding the latter.

The National Bioethics Advisory Commission, in its November 20 letter to President Clinton commenting on this experiment, rightly draws attention to the special ethical problems raised by combining human and animal cells to initiate embryonic development. On the one hand, this experiment does not create a hybrid in the sense of a being that is half human and half cow. All the nuclear genetic material comes from a human body cell; the cow egg contains some mitochondrial DNA, but this seems to be quickly taken over and directed by the human nucleus. On the other hand, proteins from the cow egg must be directing the remodeling of the chromosomes and thus the very earliest stages of development in this new being, and the ultimate effect of this is not known.

However, even if this experiment in one sense does not create a human/animal hybrid, it presents a new twist on the use of cloning to create human embryos for research purposes. Oddly, defenders of such an experiment must simultaneously argue that it is promising because it can produce genetically matched, fully human tissue for transplantation—and that it is not covered by the ban on embryo research because fusing a human nucleus and a cow egg does not really produce a human embryo.

Cows' eggs are apparently being used not to make hybrids as such, but to avoid one of the remaining practical obstacles to unlimited mass production of identical human embryos by cloning: the fact that human eggs are difficult to obtain in large numbers, and cannot be harvested in quantity without posing health dangers to women. Therefore this experiment not only poses ethical problems in its own right but could set the stage for further mistreatment of human life as an object of experimentation on a large scale. Funding for such an experiment should be, and is, banned by current law, which forbids creating a new organism from "one or more" human gametes or body cells by fertilization or cloning.

In its new letter the Commission seems very uncertain as to whether this experiment creates an embryo. But this is partly due to the Commission's own truncated approach to what constitutes an embryo. Its assumption seems to be that the new being is an embryo only if it can be proved capable of growing and developing into a new "human being," by which it means a live-born infant. But this is too narrow a standard. In many circumstances—especially those involving laboratory manipulation of new life—embryos are created in such a fatally damaged condition that they will not survive to live birth. This does not mean that they were never embryos. One might as well say that an infant born with a fatal disease, who will not survive to adulthood, was never an infant. This strange standard seems to grow out of the Commission's earlier attempts to propose that no "human cloning" has taken place so long as any human embryos created by cloning are ultimately discarded or aborted instead of being implanted to attempt a live birth. We believe the relevant question here is whether the new one-celled entity with a human nucleus begins, even for a brief time, to grow and develop as an early organism of the human species. If so, the experiment should be seen as involving the creation and destruction of human embryos and, as such, should not be funded.

Each of these new developments poses ethical questions, and none should be pursued by the federal government unless and until ethical questions have been satisfactorily answered.

A remaining question involves the other avenues for advancing stem cell research, or for advancing the medical goals to which this research is directed, without exploiting developing human beings. Last year, for example, we proposed to Congress that there may be nine promising alternatives to the use of cloning to provide stem cell lines—and eight of these seem to involve no use of embryonic stem cells at all. In the same few weeks that these embryo experiments garnered such national attention, significant advances were reported in two of these areas: The use of growth factors to help hearts grow new replacement blood vessels, and the use of stem cells from placental blood to treat leukemia and other illnesses.

It would be sad indeed if Congress's attention were to focus chiefly on those avenues of research which garner front-page news precisely because they are ethically problematic. Instead, Congress has an opportunity to use its funding power to channel medical research in ways which fully respect human life while advancing human progress. None of the new proposed experiments change in any way the ethical principle grounding current restrictions on human embryo research: In trying to serve humanity we should not support actions that are fundamentally wrong. Even a good end does not justify an evil means.

ETHICAL ISSUES IN HUMAN STEM CELL RESEARCH

REPORT AND RECOMMENDATIONS

National Bioethics Advisory Commission

THE MORAL STATUS OF EMBRYOS

To say that an entity has "moral status" is to say something both about how one should act towards that thing or person and about whether that thing or person can expect certain treatment from others. The debate about the moral status of embryos traditionally has revolved around the question of whether the embryo has the same moral status as children and adult humans do—with a right to life that may not be sacrificed by others for the benefit of society. At one end of the spectrum of attitudes is the view that the embryo is a mere cluster of cells that has no more moral status than any other collection of human cells. From this perspective, one might conclude that there are few, if any, ethical limitations on the research uses of embryos.

At the other end of the spectrum is the view that embryos should be considered in the same moral category as children or adults. According to this view, research involving the destruction of embryos is absolutely prohibited. Edmund D. Pellegrino, a professor of bioethics at Georgetown University, described this perspective in testimony given before the Commission:

> The Roman Catholic perspective . . . rejects the idea that full moral status is conferred by degrees or at some arbitrary point in development. Such arbitrariness is liable to definition more in accord with experimental need than ontological or biological reality.

Selections from "Ethical Issues in Human Stem Cell Research: Volume I Report and Recommendations of the National Bioethics Advisory Commission," September 1999.

In contrast, scholars representing other religious traditions testified that moral status varies according to the stage of development. For example, Margaret Farley, a professor of Christian ethics at Yale University, pointed out that

> There are clear disagreements among Catholics—whether moral theologians, church leaders, ordinary members of the Catholic community—on particular issues of fetal and embryo research. . . . A growing number of Catholic moral theologians, for example, do not consider the human embryo in its earliest stages . . . to constitute an individualized human entity.

Other scholars from Protestant, Jewish, and Islamic traditions noted that major strands of those traditions support a view of fetal development that does not assign full moral status to the early embryo. For example, Jewish scholars testified that the issue of the moral status of extra-corporeal embryos is not central to an assessment of the ethical acceptability of research involving ES cells. Rabbi Elliot Dorff noted that

> Genetic materials outside the uterus have no legal status in Jewish law, for they are not even a part of a human being until implanted in a woman's womb and even then, during the first 40 days of gestation, their status is "as if they were water." As a result, frozen embryos may be discarded or used for reasonable purposes, and so may stem cells be procured from them.

As a result, for some Jewish thinkers, the derivation and use of ES cells from embryos remaining after infertility treatments may be less problematic than the use of aborted fetal tissue, at least following morally unjustified abortions.

On this issue, the Commission adopted what some have described as an intermediate position, one with which many likely would agree: that the embryo merits respect as a form of human life, but not the same level of respect accorded persons. We recognize that, on such a morally contested issue, there will be strong differences of opinion. Moreover, it is unlikely that, by sheer force of argument, those with particularly strong beliefs on either side will be persuaded to change their opinions. However, there is, in our judgment, considerable value in describing some of these positions, not only to reveal some of the difficulties of resolving the issue, but to seek an appropriate set of recommendations that can reflect the many values we share as well as the moral views of those with diverse ethical commitments.

A standard approach taken by those who deny that embryos are persons with the same moral status as children and adults is to identify one or more psychological or cognitive capacities that are considered essential to personhood (and a concomitant right to life) but that embryos lack. Most commonly cited are consciousness, self-consciousness, and the ability to reason. The problem with such accounts is that they appear to be either under- or over-inclusive, depending on which capacities are invoked. For example, if one requires self-consciousness or the ability to reason as an essential condition for personhood, most very young infants will not be able to satisfy this condition. On the other hand, if sentience is regarded as the touchstone of the right to life, then nonhuman animals also possess this right.

Those who deny that embryos have the same moral status as persons might maintain that the embryo is simply too nascent a form of human life to merit the kind of respect accorded more developed humans. However, some would argue that, in the absence of an event that decisively (i.e., to everyone's satisfaction) identifies the first stage of human development—a stage at which destroying human life is morally wrong—it is not permissible to destroy embryos.

The fundamental argument of those who oppose the destruction of human embryos is that these embryos are human beings and, as such, have a right to life. The very humanness of the embryo is thus thought to confer the moral status of a person. The problem is that, for some, the premise that all human lives at any stage of their development are persons in the moral sense is not self-evident. Indeed, some believe that the premise conflates two categories of human beings: namely, beings that belong to the species *Homo sapiens,* and beings that belong to a particular moral community. According to this view, the fact that an individual is a member of the species *Homo sapiens* is not sufficient to confer upon it membership in the moral community of persons. Although it is not clear that those who advance this view are able to establish the point at which, if ever, embryos first acquire the moral status of persons, those who oppose the destruction of embryos likewise fail to establish, in a convincing manner, why society should ascribe the status of persons to human embryos.

It is not surprising that these different views on the moral status of the embryo appear difficult to resolve, given their relationship to the issues surrounding the abortion debate, a debate the philosopher Alasdair MacIntyre describes as interminable: "I do not mean by this just that such debates go on and on and on—although they do—but also that they can apparently find no terminus. There seems to be no rational way of securing moral agreement in our culture." This difficulty has led most concerned observers to search for a position that respects the moral integrity of different perspectives, but to the extent possible, focuses public policy on ethical values that may be broadly shared.

THE IMPORTANCE OF SHARED VIEWS

Once again, we are aware that the issue of the moral status of the embryo has occupied the thoughtful attention of previous bodies deliberating about fetal tissue and embryo research. Further, as already noted, we do not presume to be in a position to settle this debate, but instead have aimed to develop public policy recommendations regarding research involving the derivation and use of ES cells that are formulated in terms that people who hold differing views on the status of the embryo can accept. As Thomas Nagel argues, "In a democracy, the aim of procedures of decision should be to secure results that can be acknowledged as legitimate by as wide a portion of the citizenry as possible." In this vein, Amy Gutmann and Dennis Thompson argue that the construction of public policy on morally controversial matters should involve a "search for significant points of convergence between one's own understandings and those of citizens whose positions, taken in their more comprehensive forms, one must reject."

R. Alta Charo suggests an approach for informing policy in this area that seeks to accommodate the interests of individuals who hold conflicting views on the status of the embryo. Charo argues that the issue of moral status can be avoided altogether by addressing the proper limits of embryo research in terms of political philosophy rather than moral philosophy:

> The political analysis entails a change in focus, away from the embryo and the research and toward an ethical balance between the interests of those who oppose destroying embryos in research and those who stand to benefit from the research findings. Thus, the deeper the degree of offense to opponents and the weaker the opportunity for resorting to the political system to impose their vision, the more compelling the benefits must be to justify the funding.

In Charo's view, once one recognizes that the substantive conflict among fundamental values surrounding embryo research cannot be resolved in a manner that will satisfy all sides, the most promising approach is to seek to balance all the relevant considerations in determining whether to proceed with the research. Thus, although it is clear that embryo research would offend some people deeply, she would argue that the potential health benefits for this and future generations outweigh the pain experienced by opponents of the research.

It is, however, questionable whether Charo's analysis successfully avoids the issue of moral status. It might be argued, for example, that placing the lives of embryos in this kind of utilitarian calculus will seem appropriate only to those who presuppose that embryos do not have the status of persons. Those who believe—or who genuinely allow for the possibility—that embryos have the status of persons will regard such consequentialist grounds for destroying embryos as extremely problematic.

In our view, an appropriate approach to public policy in this arena is to develop policies that demonstrate respect for all reasonable alternative points of view and that focus, when possible, on the shared fundamental values that these divergent opinions, in their own ways, seek to affirm. This particular perspective was recommended by Patricia King in her testimony before the Commission and elsewhere. As long as the disagreement is cast strictly as one between those who think the embryo is a person with a right to life and those who think it has little or no moral status, the quest for convergence will be an elusive one. But there are grounds for supposing that this may be a misleading depiction of the conflict. Indeed, there may be a sufficiently broad consensus regarding the respect to be accorded to embryos to justify, under certain conditions, not only the research use of stem cells but also the use of embryos remaining after infertility treatments to generate ES cells.

The abortion debate offers an illustration of the complex middle ground that might be found in ethically and politically contentious areas of public policy. Philosopher Ronald Dworkin maintains that, despite their rhetoric, many who oppose abortion do not actually believe that the fetus is a person with a right to life. This is revealed, he claims, through a consideration of the exceptions that they often permit to their proposed prohibitions on abortion.

For example, some hold that abortion is morally permissible when a pregnancy is the result of rape or incest. Yet, as Dworkin comments, "[i]t would be contradictory to insist that a fetus has a right to live that is strong enough to justify prohibiting abortion even when child-birth would ruin a mother's or a family's life, but that ceases to exist when the pregnancy is the result of a sexual crime of which the fetus is, of course, wholly innocent."

The importance of reflecting on the meaning of such exceptions in the context of the research uses of embryos is that they suggest that even in an area of great moral controversy it may be possible to identify some common ground. If it is possible to find common ground in the case of elective abortions, we might be able to identify when it would be permissible in the case of destroying embryos. For example, conservatives allow such exceptions implicitly [when they] hold with liberals that very early forms of human life may sometimes be sacrificed to promote the interests of other humans. Although liberals and conservatives disagree about the range of ends for which embryonic or fetal life may ethically be sacrificed, they may be able to reach some consensus. Conservatives who accept that destroying a fetus is permissible when necessary to save a pregnant woman or spare a rape victim additional trauma might agree with liberals that it also is permissible to destroy embryos when it is necessary to save lives or prevent extreme suffering. We recognize, of course, that these cases are different, as the existence of the fetus may directly conflict with the pregnant woman's interests, while a particular *ex utero* embryo does not threaten anyone's interests. But this distinction obscures the fact that these two cases share an implicit attribution of greater value to the interests of children and adults.

We believe that the following would seem to be a reasonable statement of the kind of agreement that could be possible on this issue:

> Research that involves the destruction of embryos remaining after infertility treatments is permissible when there is good reason to believe that this destruction is necessary to develop cures for life-threatening or severely debilitating diseases and when appropriate protections and oversight are in place in order to prevent abuse.

Given the great promise of ES cell research for saving lives and alleviating suffering, such a statement would appear to be sufficient to permit, at least in certain cases, not only the use of ES cells in research, but also the use of certain embryos to generate ES cells. Some might object, however, that the benefits of the research are too uncertain to justify a comparison with the conditions under which one might make an exception to permit abortion. But the lower probability of benefits from research uses of embryos is balanced by a much higher ratio of potential lives saved relative to embryonic lives lost and by two other characteristics of the embryos used to derive ES cells: first, that they are at a much earlier stage of development than is usually true of aborted fetuses, and second, that they are about to be discarded after infertility treatment and thus have no prospect for survival even if they are not used in deriving ES cells. In our view, the potential benefits of the research outweigh the harms to embryos that are destroyed in the research process.

PETRI DISH POLITICS

Ronald Bailey

"Death to death," declares Gregory Stock, director of UCLA's Program on Medicine, Technology, and Society, at a conference on life extension. "Aging itself can be considered to be a disease," says Cynthia Kenyon, the biochemist who last year discovered genes that quadrupled the life of the nematode *C. elegans*.

"This is the first time that we can conceive of human immortality," William Haseltine, the hardheaded CEO of Human Genome Sciences Inc., the largest genomics company in the world, tells *The Washington Post*. Francis Fukuyama, the man who famously asserted that "The End of History" had arrived, declares that History is about to begin again, and its motor is biotechnology. "It is no longer clear that there is any upper limit on human life expectancy," writes Fukuyama. That, he argues, changes human nature and thus restarts History.

The biomedical revolution of the next century promises to alter our culture, our politics, and our lives. It promises to extend our life span and to enhance our mental and physical capacities. The closer those promises come to reality, however, the more they incite opposition and, in some cases, horror. And they are becoming more real by the day.

In September, Princeton University neurobiologist Joe Tsien announced that he had boosted the intelligence of mice by inserting extra copies of a gene that produces a type of receptor in brain cells; the receptor enhances

Ronald Bailey is science correspondent for *Reason* magazine and author of *ECOSCAM: The False Prophets of Ecological Apocalypse*. His articles and essays have appeared in the *Wall Street Journal*, the *Washington Post*, and *Commentary*, among others. Selections from "Petri Dish Politics," *Reason*, December 1999.

long-term memory and learning. The "smart mice" did considerably better than normal mice on a battery of six rodent intelligence tests. The mouse gene Tsien manipulated is 98 percent identical to the one found in humans. In the short term, Tsien's work could lead to drugs that will boost the memory capacities of adult humans. Over the long run, these genes might be introduced into human embryos who, once born, would have an easier time learning and retaining new information. It was the prospect of making smarter people, not just curing Alzheimer's, that made global headlines.

On the horizon are artificial chromosomes containing genes that protect against HIV, diabetes, prostate and breast cancer, and Parkinson's disease, all of which could be introduced into a developing human embryo. When born, the child would have a souped-up immune system. Even more remarkably, artificial chromosomes could be designed with "hooks" or "docking stations," so that new genetic upgrades later could be slotted into the chromosomes and expressed in adults. Artificial chromosomes could also be arranged to replicate only in somatic cells, which form regular tissues, and not in the germ cells involved in reproduction. As a result, genetically enhanced parents would not pass those enhancements on to their children; they could choose new or different enhancements for their children, or have them born without any new genetic technologies.

Already, a Vancouver company, Chromos Molecular Systems, makes a mammalian artificial chromosome that allows biotechnologists to plug in new genes just as new computer chips can be plugged into a motherboard. These artificial chromosomes, which have been developed for both mice and humans, offer exquisite control over which genes will be introduced into an organism and how they will operate.

Meanwhile, the prospect of substantially extending the human lifespan is growing, as biomedical researchers investigate promising technologies to diagnose and treat the various ways the body breaks down with age. EntreMed Inc. of Rockville, Maryland, and Cell Genesys Inc. of Foster City, California, are working to deliver a gene-based drug that will cut off a cancer's blood supply and kill it. Human Genome Sciences, also of Rockville, is developing a heart-bypass-in-a-shot using the VEGF-2 gene, which produces a protein that encourages the growth of blood vessels around blocked arteries. In Silicon Valley, Santa Clara-based Affymetrix Inc. has created a "biochip"—a silicon wafer that analyzes thousands of genes in a single test, diagnosing all sorts of diseases. Combined with the full sequence of all human genes, which will be available in a couple of years, the biochip will enable doctors to do a full genetic physical with a simple blood test.

Late last year, Geron Corp. of Menlo Park, California, announced that scientists whose work it had supported had isolated the grail of human cell biology: embryonic stem cells. These remarkable cells are capable of growing into any of the 210 types of cells found in the human body. Suffer a third-degree burn? Grow some skin cells in a petri dish for a skin graft. Heart attack? Replace the damaged tissue with made-to-order heart cells. Broken back? Fix that right up with a skein of new nerve cells.

Repairing broken bodies, extending life, and improving individuals' capabilities may sound like good things. But the promises of biomedicine increasingly attract opposition. A chorus of influential conservative intellectuals is demanding that the new technologies be crushed immediately, and many in Congress are listening. These "luddicons," as one observer has dubbed them, see in biomedicine the latest incarnation of human evil. "In the 20th century, we failed to stifle at birth the totalitarian concepts which created Nazism and Communism though we knew all along that both were morally evil—because decent men and women did not speak out in time," writes the British historian Paul Johnson in an article in the March 6, 1999, issue of *The Spectator*. "Are we going to make the same mistake with this new infant monster [biotechnology] in our midst, still puny as yet but liable, all too soon, to grow gigantic and overwhelm us?"

The most influential conservative bioethicist, Leon Kass of the University of Chicago and the American Enterprise Institute, worries both that our quest for ever-better mental and physical states is too open-ended and, contradictorily, that it is utopian. "'Enhancement' is, of course, a soft euphemism for improvement," he says, "and the idea of improvement necessarily implies a good, a better, and perhaps even best. But if previously unalterable human nature no longer can function as a standard or norm for what is regarded as good or better, how will anyone truly know what constitutes an improvement?"

Kass argues that even "modest enhancers" who say that they "merely want to improve our capacity to resist and prevent diseases, diminish our propensities for pain and suffering, decrease the likelihood of death" are deceiving themselves and us. Behind these modest goals, he says, actually lies a utopian project to achieve "nothing less than a painless, suffering-free, and, finally, immortal existence."

What particularly disturbs these conservatives is biomedicine's potential to overthrow their notion of human nature—a nature defined by suffering and death. "*Contra naturam*, the defiance of nature, used to be a sufficient argument for those who were not persuaded by *contra deum*, provoking the wrath of God," writes historian Gertrude Himmelfarb in *The Wall Street Journal.* "But what does it mean today, when we have defied, even violated, nature in so many ways, for good as well as bad?" She goes on to suggest that cloning, artificial insemination, in vitro fertilization, and even the pill might be "against nature." Himmelfarb continues, "But the ultimate question is how far we may go in defying nature without undermining our humanity. . . . What does it mean for human beings, who are defined by their mortality, to entertain, even fleetingly, even as a remote possibility, the idea of immortality?"

Himmelfarb insists that she doesn't disdain all improvement. "To raise these questions is in no way to reject science and technology or to belittle their achievements," she writes. "It is not contra naturam to invent labor-saving devices and amenities that improve the quality of life for masses of people, or medicines that conquer disease, or contrivances that allow disabled people to live, work and function normally. These enhance humanity; they do not presume to transcend it."

It is hard to see how a genetically enhanced memory, a faster mental processing speed, or a stronger immune system "undermines our humanity." After all, many full-fledged human beings already enjoy these qualities. Nor is it clear why "contrivances" that let disabled people cope with their physical problems are acceptable, while genetic cures to avoid the problems in the first place are not.

Nearly all technologies—agriculture, literacy, electric lighting, anesthesia, the pill, psychoactive drugs, television—affect human nature in the sense that they change the rhythms of human life and widen the range of behavior in which people can engage. We are no longer tribesmen living in family bands of 20, hunting and gathering on the plains of Africa. Surely there have been significant changes in human psychology as a result of the development of civilization. In fact, changing human psychology might be said to be the whole point of civilization; some anthropologists speculate that civilization is a set of social institutions that exist to tame human, especially male, violence.

Himmelfarb and Kass accuse those who favor biomedical progress of seeking immortality, as though that were a self-evident evil. But "immortality" is, in a sense, just a longer lifespan. Since 1900, lifespans worldwide have doubled, and most people think that achievement has been a great moral good. Using genetic techniques to increase human lifespans is not any different ethically from using vaccines, organ transplants, or antibiotics to achieve the same goal. Kass and Himmelfarb assert that human beings have been "defined by their mortality." But human beings are perhaps even better defined by their unending quest to overcome disease, disability, and death.

Indeed, all of the things on Himmelfarb's list of acceptable enhancements are "contra naturam." Is it not more natural to tear our meat with our hands rather than with stainless steel forks? Is it not more natural to die by the hundreds of thousands of tuberculosis, smallpox, or ebola? And is it not more natural for the lame, the blind, and old to die beneath the claws and teeth of predators? Himmelfarb does not make it clear how trying to "transcend" the dirty, nasty, brutish, and short lives of our ancestors undermines our humanity. Oh sure, a lower infant mortality rate—down from 300 or 400 deaths per 1,000 live births in the 18th century to only seven per 1,000 today—has deprived us of the chance to contemplate the tragic fleetingness of life and the poignancy of innocent death. But who among us really minds?

***

What makes us distinct and unique is not our genes but our brains and the minds they contain. Persons generally have brains that are capable of supporting enough mental activity to give rise to a mind. As one of my old philosophy professors once put it, "I have never seen a mind that was not located in fairly close proximity to a brain."

The point that brains, not genes, are the source of our uniqueness is further underscored by the fact that no one argues that natural clones, otherwise called identical twins, are the same person, even though they share an identical set of genes. They have two different brains and experience the world from two different points of view. Human brains—malleable, fluid, flexible, changing—not static genes, are the real essence of what defines us as people. We are not mere meat puppets at the mercy of our genes. In fact, with biotech it might better be said that our genes are now at the mercy of our minds.

Leon Kass is disheartened by this prospect. "We triumph over nature's unpredictabilities only to subject ourselves, tragically, to the still greater unpredictability of our capricious wills and our fickle opinions," writes Kass in the September issue of *Commentary*. In other words, he is against human freedom because he doesn't think we can handle it. Ultimately, Kass wants to preserve the "freedom" of some portion of humanity to be miserable, sick, and unhappy. But if they were truly free, would people choose to suffer or to subject their children to such suffering? Not likely.

Kass does have a point, however, when he writes in *Commentary*, "Even people who might otherwise welcome the growth of genetic knowledge and technology are worried about the coming power of geneticists, genetic engineers, and, in particular, governmental authorities armed with genetic technology."

There is a threat of government control. Some intellectuals are already succumbing to the temptation of government-supported and mandated eugenics, lest the benefits of genetic engineering be spread unequally. "Laissez-faire eugenics will emerge from the free choices of millions of parents," warns *Time* magazine columnist Robert Wright. He then concludes, "The only way to avoid Huxleyesque social stratification may be for government to get into the eugenics business."

Clearly we must be on guard against any attempts to harness this new technology to government-mandated ends. But a Brave New World of government eugenics is not an inevitable consequence of biomedical progress. It depends instead on whether we leave individuals free to make decisions about their biological futures or whether, in the name of equality or of control, we give that power to centralized bureaucracies. Huxley's world had no "laissez-faire eugenics" emerging from free choice; *Brave New World* is about a centrally planned society.

A biological future without a plan is exactly what scares critics on both the right and the left. "Though well-equipped [through biotech], we know not who we are or where we are going," Kass fearfully writes. If we know not who we are, surely advances in biotech are helping us to understand more completely who we are. As for where we are going, the fact that we don't know is why we go. Over the horizon of human discovery Kass sees a territory marked, like the maps of yore, "Here be monsters." To avoid the supposed monsters, Kass wants humanity to stay quietly at home with its old conceptions, technologies, traditions, and limited hopes.

If we use biotech to help future generations to become healthier, smarter, and perhaps even happier, have we "imposed" our wills on them? Will we have deprived them of the ability to flourish as full human beings? To answer yes to these questions is to adopt Rousseau's view of humanity as a race of happy savages who have been degraded by civilization. The fact is that previous generations have "imposed" all sorts of technologies and institutions on us. Thank goodness, because by any reasonable measure we are far freer than our ancestors. Our range of choices in work, spouses, communities, medical treatments, transportation—the list is endless—are incomparably vaster than theirs. Like earlier technologies, biotech will liberate future generations from today's limitations and offer them a much wider scope of freedom. This is the gift we will give them. Like all technologies, biotech could be abused, but using it is not, as Kass and Paul Johnson would have us believe, the same as abusing it.

Scientific facts will not resolve these issues. On the one hand, people who see human genes as the defining essence of humanity will object to stem cell research and a good deal else in the coming biotech revolution. On the other hand, people who see human beings as defined essentially by their minds will have fewer moral objections.

At a hearing earlier this year, Edward Furton, who works at the National Catholic Bioethics Center, asked the National Bioethics Advisory Commission to "please remember in your deliberations that millions of your fellow citizens hold that the human embryo is a human life worthy of the protection of the law." He added, "As a result of the tainted origin, many Americans who have deeply held moral objections to embryo destruction may choose not to receive any benefits from this new research."

No one is suggesting that people should be forced to use medicines that they find morally objectionable. Perhaps some day different treatment regimens will be available to accommodate the different values and beliefs held by patients. One can imagine one medicine for Christian Scientists (minimal recourse to antibiotics, etc.), another for Jehovah's Witnesses (no use of blood products or blood transfusions), yet another for Roman Catholics (no use of treatments derived from human embryonic stem cells), and one for those who wish to take the fullest advantage of all biomedical discoveries.

In a sense, the battle over the future of biotech—and, if Fukuyama is correct, the future of humanity—is between those who fear what humans, having eaten of the Tree of Knowledge of Good and Evil, might do with biotech and those who think that it is high time that we also eat of the Tree of Life.

REASON, FAITH, AND STEM CELLS

Michael Kinsley

Opponents of the new rules for government-funded stem-cell research are right that the rules are irrational. The rules forbid government-funded researchers to extract stem cells from human embryos but allow those researchers—on alternate Tuesdays when the wind is from the northeast and at least three members of five different review boards have dreamed of a fish—to use stem cells extracted by others.

Opponents of stem-cell research believe that "a microscopic clump of cells" (the *New York Times* description of an embryo at the stage when stem cells are removed) has the same moral claims as a fully formed human being. Proponents believe that a clump of cells has no serious moral claim compared with people who "feel want, taste grief, need friends" (Shakespeare's description of a human being). No one believes that a clump of cells is just a clump of cells in private hands but becomes a full human being in the hands of a government grantee. You don't absolve yourself of murder by hiring a hit man.

The answer to this objection (which the authors of the regulations cannot make) is: Of course it's not rational. It's a compromise between two logically irreconcilable positions. And it stretches democracy as far as it can be stretched in deference to the strongly held views of the losing side of an argument. It says: "You cannot have your way. You cannot impose the burden of your views on others. But at least you can know that your own

Michael Kinsley is editor of the online magazine *Slate.com*, a contributing writer for *Time,* and a columnist for the *Washington Post*. This essay originally appeared in the *Washington Post* on August 29, 2000.

tax dollars won't be spent directly on something you find immoral." This is quite a concession. It's more than opponents of wars, for example, are allowed.

Even the burden of this compromise is heavy on those awaiting the tremendous promise of stem-cell research. That promise has already been delayed for years by the congressional ban these new rules are designed to accommodate. The breakthroughs will be slowed by more years because of all the elaborate safeguards built in to protect those clumps of cells. Imagine being paralyzed by a spinal cord injury in your teens, watching for decades as medical treatment progresses but not quite fast enough, and knowing that it could have been faster.

In the endless right-to-life debate, compromise is difficult for pro-lifers because the strength of their side of the argument comes from its absolutism. (Unless it comes from faith, about which there can be no argument.) Absolutism is their logical trump card. If you don't protect every human being from the moment of conception, where do you draw the line? Anywhere you draw it is another irrational distinction, conferring humanity—and, possibly, life itself—on one organism and denying both to another that is nearly identical.

But absolutism is also a great weakness, because it puts you at the mercy of your own logic. Opposition to stem-cell research is the *reductio ad absurdum* of the right-to-life argument. A goldfish resembles a human being more than an embryo does. An embryo feels nothing, thinks nothing, cannot suffer, is not aware of its own existence. Embryos are destroyed routinely by the millions in the natural process of human reproduction. Yet opponents of stem-cell research would allow real people, who can suffer, to do so in service of the abstract principle that embryos are people too. If faith takes you there, fine. Reason can't.

Ronald Reagan used to play the logical trump card this way: If we don't know for sure when human life begins, we're like rescue workers after a mine explosion who don't know if anyone has survived. Shouldn't we assume there is life to be saved, rather than assuming there isn't?

The problem with this analogy is that the beginning of human life is not a factual question to which we "don't know" the answer. Biology is not going to solve this puzzle for us some day. "Human life" is a label we confer, and the uncertainty is in how we choose to define it, not in some missing bit of information. Furthermore, the definition depends on why you're asking. In the context of abortion, it doesn't matter when a fetus develops hands or feet or a heartbeat. What matters is when it develops a sense of self, an ability to suffer, or—if you go that route—an immortal soul.

And the fact that these conditions (except for the soul) don't arrive at any clear-cut moment is not the logical argument for absolutism that pro-lifers seem to think. We used to learn in high school biology that "ontogeny recapitulates phylogeny": the development of each individual human being resembles the evolution of the species. Apparently these days that is regarded as unhelpful, if not inaccurate. But even most right-to-lifers do believe in evolution and are comfortable with the idea that humanity is one end of a continuum, not a thing apart.

They are comfortable drawing a crisper line than nature does between humans and lesser beasts, and denying human rights to animals that share many human attributes. Why is it so hard for them to accept something similar about the development of an individual human being? That we each start out as something less than human, that the transformation takes place gradually, but that it's morally acceptable to draw a line somewhere other than at the very beginning. Not just acceptable, but necessary.

If faith tells you otherwise, listen. But don't mistake it for the voice of reason.

THE PIG-MAN COMETH

J. Bottum

On Thursday, October 5, it was revealed that biotechnology researchers had successfully created a hybrid of a human being and a pig. A man-pig. A pig-man. The reality is so unspeakable, the words themselves don't want to go together.

Extracting the nuclei of cells from a human fetus and inserting them into a pig's egg cells, scientists from an Australian company called Stem Cell Sciences and an American company called Biotransplant grew two of the pig-men to 32-cell embryos before destroying them. The embryos would have grown further, the scientists admitted, if they had been implanted in the womb of either a sow or a woman. Either a sow or a woman. A woman or a sow.

There has been some suggestion from the creators that their purpose in designing this human pig is to build a new race of subhuman creatures for scientific and medical use. The only intended use is to make animals, the head of Stem Cell Sciences, Peter Mountford, claimed last week, backpedaling furiously once news of the pig-man leaked out of the European Union's patent office. Since the creatures are 3 percent pig, laws against the use of people as research subjects would not apply. But since they are 97 percent human, experiments could be profitably undertaken upon them and they could be used as living meat-lockers for transplantable organs and tissue.

But then, too, there has been some suggestion that the creators' purpose is not so much to corrupt humanity as to elevate it. The creation of the pig-

J. Bottum is Books & Arts editor of the *Weekly Standard*. His essays, reviews, and poetry have appeared in the *Wall Street Journal*, *Atlantic Monthly*, *First Things*, *Commentary*, *National Review*, and *Philosophy & Literature*. This essay originally appeared in the *Weekly Standard* on October 23, 2000.

man is proof that we can overcome the genetic barriers that once prevented cross-breeding between humans and other species. At last, then, we may begin to design a new race of beings with perfections that the mere human species lacks: increased strength, enhanced beauty, extended range of life, immunity from disease. "In the extreme theoretical sense," Mountford admitted, the embryos could have been implanted into a woman to become a new kind of human—though, of course, he reassured the Australian media, something like that would be "ethically immoral, and it's not something that our company or any respectable scientist would pursue."

But what difference does it make whether the researchers' intention is to create subhumans or superhumans? Either they want to make a race of slaves, or they want to make a race of masters. And either way, it means the end of our humanity.

You can't say we weren't warned. This is the island of Dr. Moreau. This is the brave new world. This is Dr. Frankenstein's chamber. This is Dr. Jekyll's room. This is Satan's Pandemonium, the city of self-destruction the rebel angels wrought in their all-consuming pride.

But now that it has actually come—manifest, inescapable, real—there don't seem to be words that can describe its horror sufficiently to halt it. May God have mercy on us, for our modern Dr. Moreaus—our proud biotechnicians, our most advanced genetic scientists—have already announced that they will have no mercy.

It's true that Stem Cell Sciences and Biotransplant have now, under the weight of adverse publicity, decided to withdraw their European patent application and modify their American application. But they made no promise to stop their investigations into the procedure. We simply have to rely upon their sense of what is, as Mountford put it, "ethically immoral"—a sense sufficiently attenuated that they could undertake the design of the pig-man in the first place. The elimination of the human race has loomed into clear sight at last.

It used to be that even the imagination of this sort of thing existed only to underscore a moral in a story. When our ancestors heard of Vlad the Impaler's wife bathing in the blood of slaughtered virgins to keep herself beautiful, they were certain it was a bad thing. When they were told fairy tales of an old crone fattening children to suck the health from them, they knew which side they were supposed to take. When they read of Dorian Gray's purchase of eternal youth, they understood that the price he paid was his soul.

But we live at a moment in which British newspapers can report on 19 families who have created test-tube babies solely for the purpose of serving as tissue donors for their relatives—some brought to birth, some merely harvested as embryos and fetuses. A moment in which *Harper's Bazaar* can advise women to keep their faces unwrinkled by having themselves injected with fat culled from human cadavers. A moment in which the Australian philosopher Peter Singer can receive a chair at Princeton University for advocating the destruction of infants after birth if their lives are likely to be a burden. A moment in which the brains of late-term aborted babies can be vacuumed out and gleaned for stem cells.

In the midst of all this, the creation of a human-pig arrives like a thing expected. We have reached the logical end, at last. We have become the people that, once upon a time, our ancestors used fairy tales to warn their children against—and we will reap exactly the consequences those tales foretold.

Like the coming true of an old story—the discovery of the philosopher's stone, the rubbing of a magic lantern—biotechnology is delivering the most astonishing medical advances anyone has ever imagined. You and I will live for many years in youthful health: Our cancers, our senilities, our coughs, and our infirmities all swept away on the triumphant, cresting wave of science.

But our sons and our daughters will mate with the pig-men, if the pig-men will have them. And our swine-snouted grandchildren—the fruit not of our loins, but of our arrogance and our bright test tubes—will use the story of our generation to teach a moral to their frightened litters.

A CRUCIAL ELECTION FOR MEDICAL RESEARCH

Michael J. Fox

As a Parkinson's disease patient and a new American citizen, I look forward to Election Day as something momentous: It's not just the first presidential race in which I can vote (I was born in Canada). The outcome is likely to have a dramatic bearing on my prognosis—and that of millions of Americans whose lives have been touched by Parkinson's, amyotrophic lateral sclerosis, spinal cord injury, Huntington's disease, Alzheimer's disease and other devastating illnesses. That's because one question that may be decided on Tuesday is whether stem cell research—which holds the best hope of a cure for such diseases—will be permitted to go forward.

Campaign aides to George W. Bush, who has not publicly addressed the issue, stated on several occasions that a Bush administration would overturn current National Institutes of Health guidelines and ban federal funding for stem cell research. Why? Because the research, which uses human embryos discarded from fertility clinics, has become enmeshed in the politics of abortion. Mr. Bush favors a ban on stem cell research, one aide said, "because of his pro-life views."

Yet stem cell research has nothing to do with abortion. It is not the same as fetal tissue research, the federal funding of which was banned by Presidents Reagan and Bush (but has since been authorized by Congress). Stem cell work uses undifferentiated cells extracted from embryos just a few days old—embryos produced during in vitro fertilization, a process that creates

Michael J. Fox is an award-winning actor. In 2000 he founded the Michael J. Fox Foundation, which raises money for research and awareness about Parkinson's disease. This essay originally appeared in the *New York Times* on November 1, 2000.

many more fertilized eggs than are implanted in the wombs of women trying to become pregnant. Currently, more than 100,000 embryos are frozen in storage. Most of these microscopic clumps of cells are destined to be destroyed—ending any potential for life.

Their potential for saving lives, however, may be unlimited. Given the proper signal or environment, stem cells, transplanted into human tissue, can be induced to develop into brain, heart, skin, bone marrow cells—indeed, any specialized cells. The scientific research community believes that the transplanted stem cells may be able to regenerate dead or dying human tissue, reversing the progress of disease. According to Cure, a coalition of 28 groups representing patients with cancer, Parkinson's, paralysis and other maladies, "no research in recent history has offered as much hope" for cures.

Support for stem-cell research comes not just from pro-choice Democrats like Al Gore but also from Republicans who have concluded, in the words of former Senator Bob Dole, that supporting such research is "*the* pro-life position to take."

The list includes Republican senators like Strom Thurmond of South Carolina, John McCain of Arizona, Connie Mack of Florida and Pete Domenici of New Mexico. Even Senator Gordon Smith of Oregon, who the National Right to Life Committee says voted "the right way" on abortion every time last year, supports the research. His family has experience with the ravages of Parkinson's disease, and he has concluded, "Part of my pro-life ethic is to make life better for the living."

This is the real compassionate conservatism. One hopes that between now and next Tuesday, Mr. Bush will explain to those of us with debilitating diseases—indeed, to all of us—why it is more pro-life to throw away stem cells than to put them to work saving lives.

SECTION B

THE CLONING/STEM CELL DEBATE, 2001

In early 2001, two events sparked the great cloning/stem cell debate that followed: first, a number of fringe groups and fertility doctors announced their intention to clone human beings by the end of the year; and second, President Bush ordered a review of President Clinton's NIH guidelines on federal funding of embryonic stem cell research.

The debate culminated in July with a House vote to ban all human cloning, and in August with President Bush's special address to the nation on embryonic stem cells.

Almost overnight, stem cells and cloning became the dominant issues in American politics. While President Bush mulled over his decision on what to do about federal funding of embryonic stem cell research, Congress held a series of hearings on both stem cells and human cloning.

Three basic points of view emerged in the debate, though of course the complexity of the issues makes any such typology a necessary oversimplification.

PROGRESSIVES AND LIBERTARIANS

This group consists of the biotech industry, patient's advocacy groups, pro-choice liberals, and pro-progress conservatives. It favors medical research at all costs, and takes the goal of ending human suffering and disease not only as a good, but the ultimate good. It believes that scientists should be free to do their work with little or no regulation, and that individuals and parents should be free to make the decisions they deem best for themselves and their children.

In the cloning/stem cell debate, this group supports federal funding of embryonic stem cell research. It also supports so-called "therapeutic cloning," which is the cloning of embryos for research and destruction—though it often claims that such cloned embryos are not really embryos at all, since they would never be implanted to initiate a pregnancy. It generally supports a temporary ban on "reproductive cloning"—that is, the attempt to clone a child—on the grounds that human cloning is not yet safe or responsible. But it believes that most efforts to curb medical science are "religious" or "metaphysical" infringements on the separation of Church and State, even while often claiming that it is our God-given purpose to end suffering and disease and to improve the genetic condition of the human species. Proponents of this point of view include law professor Laurence Tribe, biologist Michael West, Senator Tom Harkin, and Representative James Greenwood.

ANGUISHED MODERATES

This group consists of people who seek both to champion medical progress and to maintain some moral parameters governing that progress. They accept the premise that curing disease and ending suffering is a noble, perhaps the noblest goal, but they recognize the need for some moral limits on what scientists can do—or at least on what the federal government should fund.

In the current cloning/stem cell debate, there are many kinds of anguished moderates—ranging from pro-choice liberals who believe that creating cloned embryos for research and destruction crosses an ethical line that abortion does not, to pro-life conservatives who believe that embryos outside the womb, especially if they are leftover from fertility clinics, should be used rather than "wasted." This group generally favors federal funding for stem research and at least some kind of legislative ban on human cloning. Proponents of this view include Senator Bill Frist, Senator Orrin Hatch, and political theorist Francis Fukuyama.

ANTI-BRAVE NEW WORLDERS

This group includes many who believe human embryos should never be used for medical research, and in general those who believe medical progress—and the goal of ending disease and suffering—must be governed by even higher moral obligations. Some in this group are even more radical—

believing that the modern devotion to progress itself must be called into question. They believe that while modern medicine and modern science have given us many blessings, we may be entering a period where the potential evils far outweigh the potential goods, and when leadership requires the capacity to say no and turn back. This group generally favors a permanent ban on all human cloning and an end to most or all embryonic stem cell research. Proponents of this view include Senator Sam Brownback, pro-life leader Richard Doerflinger, and bioethicist Gilbert Meilaender.

This section includes some of the major speeches, testimony, and articles from the 2001 debate: It includes a critique of the moral reasoning behind President Clinton's proposed guidelines on funding stem cell research; the testimony of those in favor and those against human cloning and embryonic stem cells (including Leon Kass, Rael, Michael West, and others); excerpts from the House debate on human cloning; and President Bush's special address to the nation and the reaction to it.

THE POINT OF A BAN

OR, HOW TO THINK ABOUT STEM CELL RESEARCH

Gilbert Meilaender

In its report *Ethical Issues in Human Stem Cell Research*, the National Bioethics Advisory Commission says the following of the congressional ban on federally funded embryo research: "In our view, the ban conflicts with several of the ethical goals of medicine, especially healing, prevention, and research." So inured have we become to such language that we fail to notice its oddity. Is it surprising that a ban should conflict with desirable goals? Or isn't that, in fact, why we sometimes need a ban—precisely to prohibit an unacceptable means to otherwise desirable ends? Taking note of this point—the oddity of NBAC's statement—should help us think about the issue of stem cell research. To explore the logic and make sense of a ban on stem cell research is my aim here. To be sure, such a ban may be persuasive chiefly for those who are concerned to affirm the dignity of the embryo, but the public debate need not be restricted to a seemingly endless argument about the embryo's status. Since many parties to the debate claim, at least, to agree that the embryo should be treated with "respect," it may be fruitful to explore other issues—in particular, the nature of moral reasoning and the background beliefs that underlie such reasoning.

I propose to take a very long way round. Our understanding of what is at stake can be sharpened if we begin not with stem cell research but with a quite different moral question.

In the memoir of his service as a Marine in the Pacific theater of World War II, historian William Manchester writes at one point:

This essay originally appeared in *Hastings Center Report*, January–February 2001.

> Biak was a key battle, because Kuzumi had made the most murderous discovery of the war. Until then the Japs had defended each island at the beach. When the beach was lost, the island was lost; surviving Nips formed for a banzai charge, dying for the emperor at the muzzles of our guns while few, if any, Americans were lost. After Biak the enemy withdrew to deep caverns. Rooting them out became a bloody business which reached its ultimate horrors in the last months of the war. You think of the lives which would have been lost in an invasion of Japan's home islands—a staggering number of American lives but millions more of Japanese—and you thank God for the atomic bomb.

Yet, one might argue—many have—that it would always be wrong to drop atomic bombs on cities, that doing so violates the rights of noncombatants. One might argue for a ban on that approach to waging war, even though in the instance cited by Manchester one can reasonably claim that such a ban would have conflicted with some of the ethical goals of statecraft: to minimize loss of life, and to seek peace and pursue it.

UTILITARIANISM OF EXTREMITY

How do we reason about such a ban in the ethics of warfare? There are, of course, different views about what is permitted in war, as there are different views on all important moral questions. But if we contemplate briefly the logic of one very widely read treatment—Michael Walzer's *Just and Unjust Wars*—we will discover that it provides a helpful window into our consideration of banning federal support for stem cell research.

Following a well-trodden path, Walzer notes that there is a kind of dualism in just war theory. It requires two different sorts of moral judgments: about when it is permissible to go to war (what Walzer calls "the theory of aggression") and about what it is permissible to do in war (which he terms "the war convention"). These are two different sorts of judgments. If we are fortunate, they will cohere for us: that is, those who have just cause for going to war will be able to win without fighting in ways that are prohibited. Because, however, these really are two different moral judgments, there are moments when we face "dilemmas of war," when it may seem, for example, that those whose cause is just cannot win unless they violate the war convention.

Confronted by such a dilemma, we might reason in several different ways. We might adopt a simple utilitarian approach; indeed, as Walzer notes, "[i]t is not hard to understand why anyone convinced of the moral urgency of victory would be impatient" with the notion of a ban on certain means to that victory. The more desirable the goals we pursue, the more tempting it will be to allow seemingly obvious utilitarian calculations to carry the day. If we take this route, the war convention provides us with rules of thumb at best. It offers some general guidelines about how to fight, which may be set aside whenever they conflict with the means required for those with just cause to win. To reason thus is in effect to conclude that the morality of war really involves only one kind of moral judgment: about when it is permissible to go to war. There is no genuine "dualism" in just war theory.

In an effort to preserve at least some sense that two different sorts of moral judgments are present, we might turn to what Walzer calls a "sliding scale." Roughly speaking, it means: Although there may be some rules that should never be violated, "the greater the injustice likely to result from my defeat, the more rules I can violate in order to avoid defeat." Some acts of war, even in a good cause, might still be wrong—if, for example, the destruction they bring is disproportionate to the good they seek to serve. But that "limit" is an essentially utilitarian one, and hence the sliding scale is simply a gradualist way of eroding the distinction between just war theory's two kinds of moral judgments. "The only kind of justice that matters is *jus ad bellum*." In short, the sliding scale is simply the timid person's avenue to utilitarian calculation.

The true alternative to such calculation seems to be a kind of moral absolutism: do justice even if the heavens fall. "To resist the slide, one must hold that the rules of war are a series of categorical and unqualified prohibitions, and that they can never rightly be violated even in order to defeat aggression." This is deontology with teeth. But it does, at least, acknowledge the force of each sort of moral judgment we make about war—what goals it would be desirable to realize, and what rights it is necessary to respect—and it permits the tension between these judgments to stand. It does not deny that winning in a just cause is often very important indeed; it simply refuses to reduce reasoning about how to fight to calculations of how best to win, and it does not gradually chip away at the rights recognized by the war convention by means of any sliding scale. In short, it acknowledges that a ban on fighting in certain ways will certainly make it more difficult to achieve the good ends sought in war, but it does not offer that fact as, in itself, an argument against such a ban. The morality of warfare involves both judgments about values to be realized and rights to be upheld. When important values cannot be realized without violating rights, it would be peculiar simply to note this fact as an argument in favor of violating rights—as if a ban on such violation were out of the question. It might be that we should do justice even if the heavens will fall, even if those values cannot then be realized or must be pursued in some slower, less certain, manner.

For such a position Walzer has considerable respect. Nevertheless, he himself adopts "an alternative doctrine that stops just short of absolutism. . . . It might be summed up in the maxim: do justice unless the heavens are (really) about to fall." This "utilitarianism of extremity" does not commit us to reasoning in terms of a sliding scale. Whether one's cause is relatively more or less just, the rules of the war convention apply with equal force, and we are not to chip away gradually at its limits. Ordinarily, a nation with just cause ought to accept defeat rather than try to win by fighting unjustly. Sometimes, however, in very special circumstances, a nation at war may face an enemy who simply "must" be defeated, whose possible victory constitutes "an ultimate threat to everything decent in our lives." The paradigmatic example of such an enemy, for Walzer, is the Nazi regime.

Confronting such an enemy, facing a defeat that threatens everything decent in human life, there might come a moment when we simply had to override the war convention and fight unjustly. This is no gradual erosion of moral limits such as the sliding scale permits. It is, rather, "a sudden breach of the convention, but only after holding out for a long time against the process of erosion." The deontological limits remain in place until the moment when we must reason in accord with a utilitarianism of extremity and override them.

How shall we recognize such a moment of supreme emergency—and, just as important, how not suppose that we face such a moment every time we are tempted to fight unjustly in a good cause? Walzer offers two criteria to help us delimit the moment, though of course criteria alone can never replace the discernment of wise men and women. It must be both strategically and morally necessary to override the war convention: no other strategy must be available to oppose the enemy, and the enemy must really constitute an ultimate threat to moral values. The moment is upon us only when we face an enemy who can be beaten in no other way, but who must be beaten. For Walzer, Britain's decision to bomb German cities—a decision made late in 1940—responded to such a moment of supreme emergency.† Civilians were targeted and the war convention overridden. Yet even in this moment of

†We should note the limits to Walzer's understanding of the supreme emergency faced by Britain: "For the truth is that the supreme emergency passed long before the British bombing reached its crescendo." Long after it was strategically necessary "the raids continued, culminating in the spring of 1945—when the war was virtually won—in a savage attack on the city of Dresden in which something like 100,000 people were killed."

supreme emergency, Walzer argues, the war convention is "overridden," not "set aside." Logically puzzling though it may be, Walzer believes that political leaders who undertake such deeds bear a burden of criminality, even though they do what they must according to a utilitarianism of extremity.

The passage from William Manchester might be thought to make such an argument from supreme emergency. "You think of the lives which would have been lost in an invasion of Japan's home islands, ... and you thank God for the atomic bomb." But Walzer believes the decision to drop the atomic bomb on Hiroshima was unjustified, and he argues that the American government did not face a moment of supreme emergency that necessitated a breach of the war convention. American policy sought from Japan an unconditional surrender, and Japanese policy was to make an invasion so costly that the Americans would prefer to negotiate a settlement. "[T]he continuation of the struggle was not something forced upon us. It had to do with our war aims. The military estimate of casualties was based not only on the belief that the Japanese would fight almost to the last man, but also on the assumption that the Americans would accept nothing less than unconditional surrender."

Since the Japanese government was not, in Walzer's view, "the moral equivalent of the Nazi regime," there was no imperative reason to demand unconditional surrender. It "should never have been asked." Of course, it would have been morally desirable—very desirable—to end the war quickly. And yes, it would have been morally desirable to end the war with a clear-cut victory. And of course it was morally desirable to minimize the loss of life. One can imagine those whose lives would have been lost had we refused to drop the bomb arguing that we might have saved them had we been less scrupulous. But for Walzer all that provided no persuasive reason to override the war convention. Hence the ban on bombing cities should never have been set aside here—good though the cause undeniably was. To say, "the ban on bombing civilians conflicts with several of the ethical goals of warfare and must therefore be set aside" would have been morally mistaken.

Two other features of Walzer's analysis need notice here before we turn to the issue of stem cell research. The first concerns his discussion of "The Dishonoring of Arthur Harris," and the second attends to the problem of nuclear deterrence. The very concept of supreme emergency assumes that, almost always, the deontological limits marked off by inviolable rights remain in place. Those limits are transgressed only in the most extreme instance of moral and strategic necessity. And they are never simply "set aside"; they are "overridden." Having been overridden, they must then be put back into place. Those who transgressed the ban and fought unjustly bear a burden of criminality. Walzer does not suppose that nation-states, especially victorious ones, could or should legally punish responsible leaders, but he does think that, after the fact, a way must be found to reinstate the overridden moral code. Thus Arthur Harris, chief of Britain's Bomber Command, who advocated bombing civilians and whose pilots carried out that terrorist policy, was the only one of Britain's top wartime commanders not rewarded after the war with a seat in the House of Lords. This "refusal to honor Harris," Walzer writes, "at least went some small distance toward re-establishing a commitment to the rules of war and the rights they protect." Supreme emergency must be a "moment." It must come to an end, and the moral law must be reacknowledged and reinstated.

To see that is to understand why one of the least successful features of Walzer's analysis of just war theory is his discussion of nuclear deterrence. The moral problem of deterrence—especially acute during the Cold War but still troubling today—is that one targets civilians, threatening almost unimaginable destruction, in order to avoid war altogether. For the many years of nuclear standoff between the United States and the Soviet Union, the posture of deterrence seemed to work (at least in the sense that nuclear weapons were used only to deter and not to fight). Walzer tries to make sense of this by suggesting that "[s]upreme emergency has become a permanent condition. Deterrence is a way of coping with that condition, and though it is a bad way, there may well be no other that is practical in a world of sovereign and suspicious states. We threaten evil in order not to do it, and the doing of it would be so terrible that the threat seems in comparison to be morally defensible." The benefits are so great that, horrifying as it is in principle, deterrence can become "easy to live with." The needed reinstatement of the moral code is deferred—indefinitely.

It is hard to find this persuasive. Having resisted any too easy transgressing of rights and limits, having confined utilitarian calculation to the moment of supreme emergency, Walzer simply settles for a permanent condition of supreme emergency. But, of course, when all moments are catastrophic, none is. In the dark of night all cats become gray, and we lose the ability to make needed and important moral distinctions.

STEM CELLS

In a recent article, Glenn McGee and Arthur Caplan argue for the moral justifiability—perhaps even obligatoriness—of stem cell research. They suggest that NBAC and other scholars (in particular, John Robertson) have been too ready to accommodate research opponents who would ban any research that involves deliberate destruction of embryos. If advocates of research cede too much ground to these opponents, they never directly confront the objection, even though they argue for moving ahead (if with caution). By contrast, McGee and Caplan argue that even if one grants the humanity and personhood of the embryo, its destruction in stem cell research is justified because this research promises to relieve incalculable suffering. Therefore, "the moral imperative of compassion . . . compels stem cell research." The "central moral issues in stem cell research" have to do, McGee and Caplan say, "with the criteria for moral sacrifices of human life." (It is instructive to note that they tend to talk not about when life may be "taken," but about when it may be "sacrificed" or "allowed to die." Clearer language would make for a clearer argument.) Thus even if one grants the personhood of the embryo, they argue, the question whether the embryo's life may be taken is still unresolved, at least for most of us. Only those who oppose all killing of any kind "can rationally oppose the destruction of an embryo solely by virtue of its status as a human person." For most of us, who do not oppose all killing as unjustified, the question becomes: "what constitutes unwarranted violence against an embryo, and for what reasons might an embryo ethically be destroyed—e.g., in the interest of saving the community?"

When, if ever, is it permissible to sacrifice a human life in service of the common good? When is such killing warranted? For McGee and Caplan, "it is clear that . . . no need is more obvious or compelling than the suffering of half the world at the hand of miserable disease. Not even the most insidious dictator could dream up a chemical war campaign as horrific as the devastation wrought by Parkinson's disease." Since it would be possible, they think, to salvage by transplantation the DNA of the embryo-to-be-destroyed, little would be lost other than easily replaceable cellular components (cytoplasm, mitochondria).[†] And they find it "difficult to imagine those who favor just war opposing a war against such suffering given the meager loss of a few cellular components."

Their argument might be summarized thus: "You think of the lives that will be lost because of serious diseases such as Parkinson's—a staggering

[†]A puzzling feature of their argument, which I cannot unpack here, has to do with this claim that the trajectory of a human life, which clearly begins with the embryo, is of little importance. As long as certain elements (DNA) are salvaged and given a new trajectory, nothing has been lost. McGee and Caplan develop their claim too briefly for one really to know what its implications are for the matter of personal identity, but, surely, they need to say far more if they are to try to make this move persuasive.

number of lives—and you thank God for stem cell research." In the face of a structurally similar argument from William Manchester and others, Walzer suggested that the United States might have changed its war aims, and that unconditional surrender was an optional goal. McGee and Caplan never consider analogous possibilities. Only unconditional surrender of Parkinson's disease will do. Progress at relieving human suffering does not seem to be an optional goal. Nor apparently is slower progress, achieved by research techniques not involving the destruction of embryos, acceptable.

Perhaps McGee and Caplan suppose that we are in something like a moment of supreme emergency. If so, they have at best made a case for moral necessity—they have identified an enemy that must be defeated. They have not yet ventured to make a case for *strategic* necessity—to show that progress cannot be made, even if more slowly, by means that do not involve destruction of embryos. Further, the case for moral necessity commits us to accepting nothing less than the eradication of all horrible diseases. Conquer one, after all, and there will be another to be conquered. Supreme emergency becomes a permanent condition, and the "sacrifice" of human lives in service of the common good and the war against suffering never comes to an end. Indeed, knowing that our actions are compelled by "the moral imperative of compassion," we act with a good conscience, bear no burden of criminality, and feel no need to find ways to reinstate the moral code we have overridden. By comparison with Walzer's analysis of just war theory, this attempt to justify stem cell research seems all too casual.

Consider a different argument about yet another issue. In a brief piece about euthanasia, written in 1990 when Jack Kevorkian had suddenly garnered attention, William F. May adopted a position on euthanasia that is not unlike Walzer's lengthier argument on the morality of war. Despite judging that the motivations behind the euthanasia movement were "understandable in an age when dying has become such an inhumanly endless business," May offered a number of reasons why acceptance of euthanasia would be bad policy. He argued that "our social policy should allow terminal patients to die but it should not regularize killing for mercy." Even the good end of relieving suffering brought on by "an inhumanly endless" process of dying did not lead May to set aside the ban on euthanasia. But he did recognize something like a moment when both moral and strategic necessity could come together in such a way as to persuade one to override that ban. "I can, to be sure, imagine rare circumstances in which I hope I would have the courage to kill for mercy—when the patient is utterly beyond human care, terminal, and in excruciating pain. . . . On the battlefield I would hope that I would have the courage to kill the sufferer with mercy." Even in such a "moment"—which can scarcely become anything like a "permanent condition"—May seems to think that the ban on killing is overridden rather than set aside and that a measure of guilt may remain. He writes that "we should not always expect the law to provide us with full protection and coverage for what, in rare circumstances, we may morally need to do. Sometimes the moral life calls us out into a no-man's-land where we cannot expect total security and protection under the law." This is the sort of argument one looks for if a ban is to be overridden.

Did NBAC do better than McGee and Caplan in offering such an argument? To some extent, it did, and although I find its approach defective, I have considerable respect for the seriousness with which it seems to have proceeded. For example, NBAC declines simply to weigh on some utilitarian balance possible relief of future suffering versus destruction of embryos. This becomes clear in its discussion of R. Alta Charo's proposal to bypass entirely the issue of the embryo's moral status. Charo suggests that we seek simply to balance deeply felt offense to some (who accept the full humanity of the embryo) over against potentially great health benefits for some future sufferers. "Thus, although it is clear that embryo research would offend some people deeply, she would argue that the potential health benefits for this and future generations outweigh the pain experienced by opponents of the research."

This "Manchesterian" argument eliminates from the outset any possibility of a ban founded on a belief that certain wrongs ought never be done. NBAC rightly notes that, at least for anyone prepared to contemplate the possibility of a ban on embryo research, Charo's recommendation must seem to be sleight of hand. "It might be argued, for example, that placing the lives of embryos in this kind of utilitarian calculus will seem appropriate only to those who presuppose that embryos do not have the status of persons." NBAC does not simply say, "You think of the suffering that will go unrelieved and the lives that will be lost without this research—and you thank God for stem cell research." It at least recognizes the force of the sort of point raised over thirty years ago by Paul Ramsey:

> I may pause here to raise the question whether a scientist has not an entirely "frivolous conscience" who, faced with the awesome technical possibility that soon human life may be created in the laboratory and then be either terminated or preserved in existence as an experiment, or, who gets up at scientific meetings and gathers to himself newspaper headlines by urging his colleagues

> to prepare for that scientific accomplishment by giving attention to the "ethical" questions it raises—if he is not at the same time, and in advance, prepared to stop the whole procedure should the "ethical finding" concerning this fact-situation turn out to be, for any serious conscience, murder. It would perhaps be better not to raise the ethical issues, than not to raise them in earnest.

NBAC's conscience is not that frivolous.

Nonetheless, it stops short of taking a ban fully seriously. Its alternative to simple utilitarian calculation seems to be a mode of reasoning analogous to Walzer's "sliding scale." Its stated aim is "to develop policies that demonstrate respect for all reasonable alternative points of view." To that end, NBAC looks for ways to express "respect" for the embryo even if not the kind or degree of respect afforded the rest of us. Hence, for example, it offers the following as "a reasonable statement of the kind of agreement that could be possible on this issue": "Research that involves the destruction of embryos remaining after infertility treatments is permissible when there is good reason to believe that this destruction is necessary to develop cures for life-threatening or severely debilitating diseases." That is, the more urgent the cause, the more potential good to be gained from this research, the more respect for the embryo must give way to the research imperative.

That this is a kind of sliding scale becomes clear when we note one of the limits recommended by NBAC. Its report supports research on spare embryos to be discarded after IVF procedures but recommends against creating embryos solely as research subjects. But this is not a limit to be respected even if the heavens will fall—or even a limit to be overridden only if the heavens are about to fall. It is a limit to be chipped away at gradually, as the little words "at this time" in the following sentence indicate: "We do not, at this time, support the federal sponsorship of research involving the creation of embryos solely for research purposes. However, we recognize that in the future, scientific evidence and public support for this type of stem cell research may be sufficient in order to proceed."

This is a kind of "proceed with caution" view. One suspects that the chief "limit" to research discerned by NBAC involves not so much the status of the embryo as the status of "public support."[†] There is no sense here of a limit that could be overridden—if at all—only in a moment of supreme emergency, which

[†]I do not wish to deny the obvious fact that a public commission such as NBAC must pay attention to and measure public support when it makes recommendations. Nevertheless, if one accepts a research ban as one choiceworthy moral option, then one must be open to the possibility that NBAC's responsibility might be to marshal public support for such a ban.

overriding would involve a burden of criminality, and which limit would somehow have to be reinstated after the fact. Such an argument, if it could be made persuasively, would be a very strong expression of respect for embryos. NBAC does much less, however. From one perspective, in fact, perhaps NBAC's cautious sliding scale shows less respect for embryos than McGee and Caplan's "full speed ahead" approach, since one can read McGee and Caplan as justifying stem cell research with a kind of "supreme emergency as a permanent condition" argument. While I doubt that it really makes sense to posit such a permanent condition of supreme emergency, the attempt does at least acknowledge that nothing less than such extreme circumstances could even claim to justify embryo research. The more judicious, "at this time" approach of NBAC promises, by contrast, a kind of relentless "progress" in what is allowed. It is not really prepared ever to stop. It cannot contemplate or make sense of a ban.

ENDS AND MEANS

Perhaps we can understand, then, why some critics of stem cell research would not be persuaded by moral reasoning that uses simple utilitarian calculation, applies a "sliding scale," or appeals to "supreme emergency as a permanent condition." If we are among the unpersuaded, we are left to contemplate seriously a ban. To do that, however, may compel us to think also about the background beliefs—metaphysical and religious in character—that undergird all our moral reflection. In particular, we will be forced to ponder the degree to which relief of suffering has acquired the status of trump in our moral reasoning.

Why might one, even while granting the enormous benefits to be gained from stem cell research, be prepared to contemplate a ban on research that requires the destruction of embryos? How must one think for such a ban to make sense? Clearly, no ban can make sense if we say with McGee and Caplan that "no need is more obvious or compelling than the suffering of half the world at the hand of miserable disease." Nor could any ban make sense in the context of a search, such as NBAC's, for a public policy "consensus" that, while taking objections seriously, will always permit research to proceed. Indeed, despite NBAC's serious attempt to be fair-minded, its understanding of consensus ultimately excludes from consideration precisely those who might be willing to think in terms of a ban.

The very notion of a ban can make sense only if we consider that the fundamental moral question—for a community as for an individual—is *how* we live, not *how long*. If we act simply for the sake of future good, the day will come when those good effects reach an end—which is not a telos, but simply an end. We will have done evil in the present for a future good that does not come to pass.

In his meditations to himself, Marcus Aurelius writes: "Another [prays] thus: How shall I not lose my little son? Thou thus: How shall I not be afraid to lose him?" That is, how shall I not be afraid if the alternative to losing him is doing wrong? In our tradition this emphasis on means over ends, on how rather than how long we live, has been grounded not only in such Stoic thought but also and primarily in Jewish and Christian belief.

It has provided the moral background that makes sense of doing justice even if the heavens are about to fall.

One who looks on life this way need not, of course, suppose that beneficence is unimportant or that relief of suffering is of little consequence. Weighty as such values are, however, they have no automatic moral trump. To appreciate this, we can consider passages from two twentieth century thinkers for whom it was clear that the most important moral question was how we live. In *The Screwtape Letters* C.S. Lewis created a series of letters from a senior devil to a junior tempter on the subject of how to tempt a mortal—with instructions that invert the moral world by inviting us to look at things from the perspective of Satan (for whom God must be "the Enemy"). So, for example, Screwtape advises Wormwood about the attitude toward time that he ought to cultivate in his patient:

> [N]early all vices are rooted in the Future. Gratitude looks to the Past and love to the Present; fear, avarice, lust, and ambition look ahead. . . . [The Enemy] does not want men to give the Future their hearts, to place their treasure in it. We do. . . . [W]e want a man hagridden by the Future—haunted by visions of an imminent heaven or hell upon earth—ready to break the Enemy's commands in the Present if by so doing we make him think he can attain the one or avert the other.

Likewise, reflecting upon "the ethics of genetic control," Paul Ramsey noted the relatively greater importance of an "ethics of means" for religious thinkers:

> Anyone who intends the world as a Christian or as a Jew knows along his pulses that he is not bound to succeed in preventing genetic deterioration, any more than he would be bound to retard entropy, or prevent planets from colliding with this earth or the sun from cooling. He is not under the necessity of

> ensuring that those who come after us will be like us, any more than he is bound to ensure that there will be those like us to come after us. He knows no such absolute command of nature or of nature's God. This does not mean that he will do nothing. But it does mean that as he goes about the urgent business of doing his duty in regard to future generations, he will not begin with the desired end and deduce his obligation exclusively from this end. . . . And he will know in advance that any person, or any society or age, expecting ultimate success where ultimate success is not to be reached, is peculiarly apt to devise extreme and morally illegitimate means for getting there.

My aim is not to inject religious beliefs into public discussion of stem cell research. On the contrary, my point is that such beliefs are already there. To see clearly the kind of background beliefs which might make a ban on stem cell research seem reasonable is also to realize that something like a religious vision of the human is at work in arguments *for* such research. Precisely insofar as a ban is not really an option, insofar as proponents of a ban cannot possibly be included in any proposed consensus, the argument for research is that we—human beings—bear ultimate responsibility for overcoming suffering and conquering disease. We know along our pulses that we are, in fact, obligated to succeed, compelled to ensure that future generations not endure suffering that we might have relieved. Possible future benefits so bind our consciences that we are carried along by an argument we might well reject in, say, the ethics of warfare. "You think of the suffering that will go unrelieved and the lives that will be lost without this research and you thank God for stem cell research."

It is quite true, of course, that a ban on stem cell research requiring destruction of embryos would mean that future sufferers could say to us: "You might have made more rapid progress. You might have helped me." To consider how we should respond to them is to contemplate the moral point of a ban: "Perhaps we could have helped you, but only by pretending that our responsibility to do good is godlike, that it knows no limit. Only by supposing, as modernity has taught us, that suffering has no point other than to be overcome by human will and technical mastery—that compassion means not a readiness to suffer with others but a determination always to oppose suffering as an affront to our humanity. We could have helped you only by destroying in the present the sort of world in which both we and you want to live—a world in which justice is done now, not permanently mortgaged in service of future good. Only, in short, by pretending to be something other than the human beings we are."

WHY PRO-LIFERS ARE MISSING THE POINT

Charles Krauthammer

The abortion wars are on again. No, abortion is not about to be outlawed. There will be no overturning of *Roe v. Wade*. In America, this battle is fought, peculiarly, not at the center but at the periphery. The new President repeals the former President's directive allowing funding for abortion counseling overseas. He orders a safety review of RU 486, the so-called abortion pill. He then expresses himself on perhaps the most peripheral issue of all: research that relies on fetal tissue. Bush opposes such research, and has asked the Department of Health and Human Services to study whether federal funding for it should be banned. Now, there may be good reason to pause before opening wide the doors to this kind of research—but not for the reasons being advanced by opponents of abortion. The real problem is not where the cells come from, but where they are going.

At immediate issue are "stem cells," cells often taken from the very earliest embryo. Because they are potentially capable of developing into any kind of cell, they may help cure an array of intractable diseases. Pro-life forces find the procedure ethically impermissible, because removing the cells kills the embryo. Moreover, they argue, harvesting this biological treasure will encourage the manufacture of human embryos for precisely this utilitarian purpose.

But their arguments fail. First, stem cells are usually taken from embryos produced for in-vitro fertilization or from aborted fetuses. Both procedures are legal. They produce cells of incalculable value that would otherwise be discarded. Why not derive human benefit from them? Second, the National

This essay originally appeared in *Time* magazine on February 12, 2001.

Institutes of Health guidelines issued last August take away any incentive to abort or otherwise produce embryos just for their useful parts: no payment for embryos and no dedication of embryonic cells for specific recipients (say, for injection into a sick family member). Finally, there is the potential benefit. Because embryonic stem cells can theoretically develop into any cell type in the body, they could cure all kinds of diseases, such as Parkinson's, diabetes and Alzheimer's. Will it work? We can't know without the research.

One can admire pro-lifers for trying to prevent science from turning human embryos into tissue factories. But theirs is a rearguard action. The benefits of such research will soon become apparent. Stem cells are now being injected into monkeys with a Lou Gehrig's-like disease. Human trials will undoubtedly follow. Those resisting this research will find themselves outflanked politically, as the stampede of the incurably sick and their loved ones rolls through Congress demanding research and treatment. The resisters will also find themselves outflanked morally when the amount of human suffering that stem cells might alleviate is weighed against the small risk of increasing the number of embryos that do not see life.

In their desire to keep the embryo inviolable, opponents are missing the main moral issue. The real problem with research that manipulates early embryonic cells—whether derived from fetal tissue or from adult cells rejuvenated through cloning—is not the cells' origin but their destiny. What really ought to give us pause about research that harnesses the fantastic powers of primitive cells to develop into entire organs and even organisms is what monsters we will soon be capable of creating.

In 1998, Massachusetts scientists injected a human nucleus into a cow egg. The resulting embryo, destroyed early, appeared to be producing human protein, but we have no idea what kind of grotesque hybrid entity would come out of such a marriage. Last October, the first primate containing genes from another species—a monkey with a jellyfish gene—was born. Monkeys today. Tomorrow humans.

Just last month Britain legalized embryonic stem-cell research. But it did not stop there. Parliament also permitted "therapeutic" human cloning. That means that you cannot grow your clone in a uterus to produce a copy of yourself, but you can grow it in a test tube to produce organs as spare parts. Anyone who believes that such lines will not be crossed is living on the moon.

The heart of the problem is this: It took Nature 3 billion years of evolution to produce cells that have the awesome power to develop autonomously, through staggeringly complex chemical reactions, into anything from a kidney cell to a full thinking human being. We are about to harness that power for crude human purposes.

What will our purposes be? Of course there will be great medical benefits. They will seduce us into forging bravely, recklessly ahead. But just around the corner lies the logical by-product of such research: the hybrid human-animal species, the partly developed human bodies for use as parts, and other grotesqueries as yet unimagined. That is what ought to be giving us pause: not where we took these magnificent cells from but where they are taking us.

THE POLITICS OF STEM CELLS

Wesley J. Smith

Stem cells are undifferentiated "master cells" in the body that can develop into differentiated tissues, such as bone, muscle, nerve, or skin. Stem cell research may lead to exponential improvements in the treatment of many terminal and debilitating conditions, from cancer to Parkinson's to Alzheimer's to diabetes to heart disease. Indeed, breakthroughs in stem cell research reported just in the last six months take one's breath away:

- Italian scientists have generated muscle tissue using rat stem cells, a discovery that may have significant implications for organ transplant therapy.
- University of South Florida researchers report that rats genetically engineered to have strokes were injected with rat stem cells that "integrated seamlessly into the surrounding brain tissue, maturing into the type of cell appropriate for that area of the brain." The potential for stem cell treatments to alleviate stroke symptoms such as slurred speech and dizziness—therapy that would not require surgery—has the potential to dramatically improve the treatment of many neurological diseases.
- The group of scientists who achieved worldwide fame for cloning Dolly the sheep have successfully created heart tissue using cow stem cells. The experiment demonstrated that stem cells could be trans-

Wesley J. Smith is the author of *Culture of Death: The Assault on Medical Ethics in America* and an attorney for the International Anti-Euthanasia Task Force. He is the co-author of four books with Ralph Nader and of numerous articles on corporate irresponsibility and medical ethics. This essay originally appeared in the *Weekly Standard* on March 26, 2001.

formed into differentiated bodily tissues, offering great impetus to further research.

- Scientists at Enzo Biochem, Inc., inserted anti-HIV genes into human stem cells. The stem cells survived, grew, and developed into a type of white blood cell that is affected adversely by HIV infection. In the laboratory, these treated cells blocked HIV growth. The next step is human trials, in which stem cell therapy will be attempted using bone marrow transplantation techniques currently effective in the treatment of some cancers.

What will surprise many people is that *none* of these remarkable achievements relied on the use of stem cells from embryos or the products of abortion. Indeed, all of these experiments involved *adult stem cells* or undifferentiated stem cells obtained from other non-embryo sources. The rat muscle tissue in the first example was generated using adult rat brain cells. The brain tissue generated in the Florida research was obtained using human stem cells found in umbilical cord blood—material usually discarded after birth and a potentially inexhaustible source of stem cells, since 4 million babies are born in the United States alone each year. Dolly's creators obtained cow heart tissue by reprogramming adult cow skin tissue back into its primordial stem cell state and thence to cardiac cells. The exciting HIV experiments were conducted using stem cells found in the patients' own bone marrow, spleen, or blood.

The opportunities for developing successful therapies from stem cells that do not require the destruction of human embryos should be very big news. But where are the headlines? These and other successful experiments have been all but drowned out by breathless stories extolling the miraculous potential of embryonic stem cell research. How many readers are aware, for example, that French doctors recently transformed a heart patient's own thigh muscle into contracting muscle cells? When these cells were injected into the patient's damaged heart, they thrived and, in association with bypass surgery, substantially improved the patient's heartbeat. Such research is now on the fast track, offering great hope for cardiac patients everywhere.

With all of the hype surrounding embryo research, it is important to note that embryo stem cell research—and its first cousin, fetal tissue experiments—may not actually produce the therapeutic benefits its supporters have told us to anticipate. Such worries are not mere speculation. The March 8, 2001, *New England Journal of Medicine* reported tragic side effects from an experiment involving the insertion of fetal brain cells into the brains of Parkinson's disease patients. The patients thus treated showed modest if any overall benefits by comparison with a control group who underwent "sham surgeries" without receiving fetal tissue. But over time, some 15 percent of the patients who had received the transplants experienced dramatic over-production of a chemical in the brain that controls movement. The results, in the words of one disheartened researcher, were "utterly devastating," with the unfortunate patients exhibiting permanent uncontrollable movements: writhing, twisting, head-jerking, arm flailing, and constant chewing. One man was so badly affected he no longer can eat, requiring the insertion of a feeding tube.

While some studies using stem cells culled from embryos to treat Parkinson's type symptoms in mice have been encouraging, grafts of fetal and embryonic tissue may provoke the body's immune response, leading to rejection of the tissue and potentially death, since once the cells are injected they cannot be extracted. Even more alarming, a May 1996 *Neurology* article disclosed a patient's death caused by an experiment in China in which fetal nerve cells and embryo cells were transplanted into a human Parkinson's patient. After briefly improving, the patient died unexpectedly. His autopsy showed that the tissue graft had failed to generate new nerve cells to treat his disease as had been hoped. Worse, the man's death was caused by the unexpected growth of bone, skin, and hair in his brain, material the authors theorized resulted from the transformation of undifferentiated stem cells into non-neural, and therefore deadly, tissues.

Even some of the most enthusiastic boosters of embryo stem cell research see trouble ahead. For example, University of Pennsylvania bioethicist Glenn McGee admitted to *Technology Review,* a Massachusetts Institute of Technology publication, "The emerging truth in the lab is that pluripotent stem cells are hard to rein in. The potential that they would explode into a cancerous mass after a stem cell transplant might turn out to be the Pandora's box of stem cell research." Thus, it could be that adult tissue-specific stem cells are actually safer than their counterparts culled from embryos since, being extracted from mature cells, they may not exhibit the propensity for uncontrolled differentiation.

These concerns arise just as the long-time ban on using federal funds for research that destroys human embryos is under renewed scrutiny. That long-standing ban was effectively reinterpreted out of existence in the waning months of the Clinton administration, and the National Institutes of Health are currently accepting grant proposals for research using embryos originally created for in vitro fertilization but now deemed "in excess of clinical need."

The new administration is taking a long, hard look at the policy; during the campaign, George W. Bush declared his opposition to research that involved destroying human embryos.

All of this raises intriguing questions: Why is federal funding for embryo and fetal research pushed so hard and so publicly—while adult stem cell and other alternative therapies are damned with faint praise? Why do the media applaud fetal stem cell experiments and provide klieg-light coverage of stories promoting the use of embryos, while they mention uncontroversial research not requiring the destruction of human life as an afterthought, if that? Indeed, why do some scientists assert that alternative stem cell research offers but uncertain hope, while they promote embryo and fetal tissue research as the keys to the Promised Land?

I suggest three answers: celebrities, abortion, and eugenics.

In a society that has often denigrated its true heroes, the only people who now stand head above the clouds are figures from the world of entertainment. Increasingly, these celebrities are using their power to promote public policies. They know that their participation can define issues and shape the debate by attracting media coverage, generating fan support, and, most important, stimulating a Pavlovian response in politicians.

Three high-powered celebrities have weighed in recently in the stem cell controversy, each promoting full federal funding of embryo research: the popular Michael J. Fox, stricken at a tragically young age with Parkinson's disease; the television icon Mary Tyler Moore, a diabetes patient; and actor Christopher Reeve, paralyzed from the neck down in an equestrian accident. With such kiloton star power favoring federal funding of embryo research, promoters of research relying on adult stem cells and other alternative sources, along with those opposed to the destruction of embryos on ethical grounds, have been reduced to background noise or, worse, made to look heartless by denying these celebrities medical breakthroughs they need.

At a deeper level, just as in the nineteenth century many national issues led back to slavery, today numerous public policy disputes lead ultimately to abortion. The controversy over destroying human embryos to obtain their stem cells has brought an outcry from the pro-life movement, which views human life as sacred from the moment of conception. This has led to reflexive support for embryo research by many pro-choicers, who have seized on the issue as a way to further their depiction of pro-life forces as caring little about people once they are born. Thus the embryo stem cell debate offers abortion rights advocates a "two-fer": It furthers their primary political goal of isolating and marginalizing pro-lifers, and it enables them to seize the PR high ground by "compassionately" pressing for research that offers hope against debilitating diseases. To acknowledge the tremendous potential of adult stem cell research would interfere with this political pincer movement.

Finally, in my view, the ultimate purpose of promoting federal funding for embryo experiments over adult stem cell research—particularly among many in the bioethics movement—is to open the door to the eugenic manipulation of the human genome. Once embryos can be exploited for their stem cells to promote human welfare, what is to stop scientists from manipulating embryos to control and direct human evolution—equally for the purpose of improving the human future?

Indeed, some of those who signed a recent open letter to President Bush urging an end to the ban on federal funding for human embryo research were scientists and bioethicists well known as favoring eugenics. For example, James D. Watson, a co-discoverer of the DNA helix, has written that newborns should not be considered "alive" for three days, to permit genetic screening. Newborns who fail to pass genetic muster should be discarded—much as the ancient Romans left unwanted babies outdoors to die of exposure. Another co-author of this letter, Michael West, head of the for-profit research company Advanced Cell Technology, proposes permitting human cloning as a way to obtain genetically matched stem cells for transplants, which might overcome the problem of tissue rejection in embryo stem cell therapy. Not coincidentally, many neo-eugenicists in the bioethics and science communities view cloning as a prime vehicle for directing the eugenic manipulation of human evolution.

All of this will come to a head in the coming weeks and months. Some recent news stories indicate that Health and Human Services secretary Tommy Thompson may be troubled by a federal ban on embryo stem cell research and thus inclined to retain the Clinton administration's funding policy. But why go down that controversial path, when adult stem cells and alternative sources offer such tremendous hope for treating every malady that research using embryos and fetal tissue seeks to ameliorate? Instead of turning this important field of medical research into another battlefield in America's never-ending culture war (the first lawsuit has already been filed to prevent federal funding), why not focus our public resources with laser-like intensity on the incredible potential of adult and alternative sources of stem cells?

TESTIMONY

U.S. House of Representatives

Rael

The conservative, orthodox, fanatic traditional religions have always tried to keep humanity in a primitive stage of darkness. It is easy to see that in countries like Afghanistan which are back to the middle ages due to a fanatic Moslem government.

But this was also true in occidental powers. The first medical doctors who tried to study the human by opening cadavers were excommunicated by the Catholic Church. It was considered a sin to try to unveil the mystery of the creation of god. . . So were the first antibiotics, blood transfusions, vaccines, surgeries, contraception, organ transplants. . . religious fanatics were always saying that "it's a sin to go against the will of god. . . If somebody is sick, let him die, his life is in god's hands."

If our civilization would have respected these primitive ideas from dark ages, we would all die around 35, and 9 out of 10 babies born would die in their first 2 years.

Traditional religions have always been against scientific progress. They were against the steam engine, electricity, airplanes, cars, radio, television, etc. . . If we had listened to them we would still have horses and carts and candles. . .

Twenty-two years ago they were against IVF, talking about monsters, Frankenstein and playing god, and now IVF is well accepted, performed

Rael is the head of the Raelian cult. He believes he was contacted in 1973 by aliens and that human cloning and genetic engineering are the key to becoming like our extraterrestrial creators. Selections from testimony before the U.S. House of Representatives Committee on Energy and Commerce, Subcommittee on Oversight and Investigations, March 28, 2001.

every day by thousands and helping happy families with fertility problems to have babies.

Today human cloning will help other families to have children, and again they are against it. It will also help to cure numerous diseases, will help us live a lot longer, and finally will help us reach, in the future, eternal life.

Nothing should stop science, which should be 100 percent free.

Ethical committees are unnecessary and dangerous as they give power to conservative, obscurantist forces, which are guided only by traditional religious powers.

As well as there should be a complete separation of state and religion, there should also be a complete separation of science and state, or science and religion.

If there was an ethical committee when antibiotics, blood transfusions and vaccines were discovered it would have certainly been possible that these technologies would have been forbidden. You can imagine the poor health the world would be in today. . .

Ethical committees should be necessary when a deadly technology is making the production of weapons of mass destruction possible. . . And to my knowledge there are no ethical committees concerning nuclear, chemical or biological weapons. These things are created to kill millions of people and possibly destroy all life on earth. Cloning is a pro-life technology, a technology made to give birth to babies!

The first benefit of human cloning is to make it possible for couples who cannot have children using other existing techniques to have babies inheriting genetic traits from one of their parents. They can be unfertile heterosexual couples or gay couples.

The second benefit is for families who lose a child due to crime, accident or disease to have the same child brought back to life.

All conservative "pro-life" groups always talk about "the right of the unborn," but in this case we must talk about protecting the rights of the "unreborn." As cloning technology makes this possible, why should we accept the accidental death of a beloved child, when we can bring this very child back to life?

People who are opposed to it are always influenced by a terrible Judeo-Christian education . . . the same as those who could have made antibiotics, vaccines, transfusions, surgery and organ transplants forbidden. Their main objections are:

1. "It is an unsafe technology, which is not advanced enough": the best way to develop this technology like all other technologies is by doing

it. The first surgeries, organ transplants, and IVF were unsuccessful. But by doing it, scientists were able to develop their expertise.

2. "It will create monsters": a percentage of "normally" (sexually) conceived babies are born "monsters" or with genetic faults . . . Would you make a law against making babies sexually through the "natural" way because of these problems? Of course not and the percentage amongst cloned babies will be lower as they will be more precisely scrutinized from the first days after conception.
3. "The children made by cloning will have a terrible life being looked at as abnormal people": not more than IVF conceived children, or twins, or physically handicapped children or gay or colored people. It is not the problem of the children themselves, but the responsibility of the society to educate the public to respect the differences, all the differences, between human beings.
4. "It is against biodiversity to create cloned children, creating identical people": we have already on earth millions of twins and this is not a problem. The conception through cloning will always be used by a limited number of people and that will not affect biodiversity. But even if you imagine 6 billion human beings being cloned, the biodiversity is still the same as we still would be 6 billion different people!
5. "Cloned children will not be exactly the same": so what is the problem? As long as the families are informed about this, (and they are) there is absolutely no problem with that. People against cloning keep saying "it is terrible they will be the same" and then suddenly they argue "they will not be exactly the same" . . . So what is the problem?
6. "Cloned children will have terrible psychological problems being created to replace dead babies, but they will never be exactly the same": loving families who lose a child and want to have him back through cloning have so much love for this child, a child they hope so much to have back, that I cannot imagine a child being loved more. More than other families, those who lose a child due to disease or accident or crime have learned so much about how life is fragile, that they will protect and care for these children much more than "normal" families. And a good education is to accept that your children are not exactly what you would want them to be. "Normal" families experience that every day when a father who is a medical doctor sees his child only interested by music or painting, as the father was dreaming to have his son become a doctor like he is. . . Real love is ac-

cepting the differences, and that includes the differences between the image you have of your child and who he really is.

7. "Human cloning is unnatural": we already answered to this objection. Nor are blood transfusions, antibiotics, organ transplants, vaccinations, surgery, etc . . . If we let "mother nature" work we would be all dead around 45 and 9 out of ten babies would die as infants . . .
8. "Only God can create life": this is pure belief, and religious people who are against human cloning have the right and the freedom not to do it, as they can refuse blood transfusion, organ transplant, surgeries, antibiotics, contraception, abortion, etc., but those who decide to do it should be respected as well.

These are the most frequent objections to human cloning, but we should also consider the advantages in the middle and long term aspects of human cloning for a non-fanatically religious society.

Human cloning will help cure numerous if not all diseases. It will also make [it] possible to create a genetic bank where you will be able, if you need an organ transplant, to have it. Not by creating babies to take replacement organs from them . . . but by preserving stem cells of your body in very early stage embryos of yourself and develop[ing] in vitro only the organs you need in case of disease or accident. . . [T]hose opposed for religious reasons . . . should be free not to use it.

Finally, in the more long term, human cloning will make [it] possible—when Accelerated Growth Process [is] discovered to clone an adult copy of ourself directly just before we die and when Brain Data Transfer [is] discovered, to transfer, or download (or upload) our memory and personality in this new young body for a new long life. The progress of humanity will be exponential at this level. Presently, when a scientist is at the top of his art, he starts to age and dies. We can imagine Einstein, Newton and Leonardo Da Vinci still alive and working together . . . the discoveries they could do would be unlimited. And the same for artists like Mozart, Beethoven and Bach being still alive.

Not only should human cloning be allowed for the good of today's people, but even more for future generations who will remember your historical decision forever.

That's why I chose America to create the first Human Cloning company, because it is the country of individual freedom and science on earth. Thanks to the U.S. system, which should be a model for the whole world, and special thanks to the Supreme Court, I am confident that the right to clone yourself as an individual freedom, guaranteed by the great U.S. Constitution, will be protected.

AGAINST HUMAN CLONING

J. Bottum

Last week, the Brownback-Weldon bill to prohibit human cloning was introduced on Capitol Hill. And the arguments against it are . . . well, as it turns out, there really aren't many arguments against a ban on manufacturing human beings like gingerbread men from a cookie cutter.

It's true, of course, that some propositions resembling arguments for cloning have been advanced in recent years. But under scrutiny, these ostensible arguments quickly dissolve into a fog of vague, unfocused *feelings* about science, sex, and the human condition.

Take, for example, the claim that to prohibit cloning would be to prevent a grief-stricken mother and father from replacing their dead daughter with a new, genetically identical daughter who will somehow erase the loss of their first daughter. You don't have to delve very far into philosophical questions of identity and existence to realize that the notion is so confused and self-contradictory, it won't even bear the weight of its own expression. But the point of invoking those grieving parents is not to present an argument. The point is to express a feeling: Death ought not to sting, the grave should not have the victory, the ones we love must come back to life. And so cloning enthusiasts look to science—as to a god—to wipe away our tears, to assuage the eternal pity, and to console human grief.

Or take, for another example, the claim that a ban on human cloning would be a blow against *Roe* v. *Wade.* Some antiabortion activists do make this argument. They say everything bad begins with a disrespect for human life: The unfettered right to abortion grants us a Promethean power of life

This essay originally appeared in the *Weekly Standard* on May 7, 2001.

and death over our unborn offspring that naturally leads to practices like cloning. Thus, the argument goes, we can succeed in banning cloning only by winning—today—the battle over abortion. Many supporters of cloning actually make the same argument, although they run it in reverse to frighten off liberal Democrats: A ban on cloning, they warn, would mean the loss of "a woman's right to choose"; America can thus guarantee the full abortion license only by allowing cloning to proceed unhindered.

Our fellow pro-lifers may well be right that there is an underlying logic linking these issues. But the truth is—and this is the vital political point—we can ban cloning without touching *Roe* v. *Wade.* Indeed, the debate over cloning shouldn't be forced back into the well-worn grooves of the abortion debate. The issue of cloning offers the possibility of some interesting realignments in American politics. This is an issue, after all, on which radical environmentalists and religious evangelicals find themselves in agreement—which would be impossible if the right-to-choose equals right-to-clone argument were definitive. But, then, this was never meant to be a genuine argument. It is meant instead to express a feeling—a feeling that radical individualism, sexual liberation, and modern science have all somehow combined to bring us to this point, and to reject any piece of it now, even the reproduction of human beings by cloning, is to return to the Dark Ages.

And take, for a final example, the claim that a law against cloning human beings will make us forfeit potential advances in medicine. Who could be opposed to experiments that might lead to a cure for cancer, a fully compatible liver for transplanting, a genetically engineered solution to diabetes? But examined more closely, the hoped-for medical advances turn out to be merely examples of things that researchers promise they will try to find, if only we leave them alone to play with human cloning as much as they like.

The manipulation of stem cells obtained from cloned embryos is asserted to be necessary for the desired medical breakthroughs. And the use of these putative therapeutic miracles in pro-cloning arguments seems to have survived unscathed the recent evidence that it is possible to obtain the required stem cells not from embryos but from adults' blood, bone marrow, brain tissue, and even fat cells. It has survived unscathed, for that matter, the disastrous initial results of stem-cell treatment (in which the cells, derived from embryos, went wild and began producing not merely brain tissue but other tissue as well when introduced into the brains of some of their new hosts).

But, then, the promise of unlimited medical advance was never really an argument for keeping cloning legal. It is a feeling, a sentiment, masquerading as an argument—and perhaps the most insidious of them all. A vague belief in the capacity of human beings to obtain any end through beneficent science has oddly joined a vague belief in the *in*capacity of human beings to halt the march of science or decide what those ends should be. Once we add in the thousands of university laboratories anxious for the acclaim of scientific breakthroughs and the dozens of large pharmaceutical companies hungry for new technologies, the use of cloning simply feels like the future: unavoidable, inexorable, and predetermined. As well oppose the rising of tomorrow's sun, we are counseled, as try to halt the arrival of human cloning.

Yet halt it we can, and should—for reasons compellingly presented by such thinkers as Leon Kass and Gilbert Meilaender. Those reasons range from the extraordinarily high incidence of deformity among cloned animals, to the familial confusion that will be engendered by reproducing oneself as one's own child, to the likely psychological damage to the person created by cloning, and, most fundamentally, to the fact that moving from the begetting of our children to the manufacture of our descendants is a radical and perhaps irreparable dehumanization.

American politics being what it is, there will be an attempt to find a "compromise" on this issue, as there was when Congress last considered it in 1998. The favored form of compromise prohibits "reproductive" cloning while allowing "therapeutic" cloning to continue unabated. But a ban solely on reproductive uses only *looks* like a compromise. It's actually a victory for the pro-cloning forces—and everyone opposed to the onslaught of human cloning must reject it out of hand. For what this "compromise" would mean is a license to practice all the cloning a scientist may desire, while vainly attempting to prevent the end toward which that practice clearly aims: the live birth of cloned human beings.

Part of the problem is the question of intention. Since all embryonic clones are made in the same way, we cannot know the reason for which an embryo was created until it is either destroyed in research or implanted in a womb. Of course, once it has been implanted, a law against reproductive cloning would clearly have been violated. But there is at that point no possible redress, short of forced abortions or a federal pregnancy police determining how each pregnancy in America came about.

Then, too, there is the problem of the status of the embryos created by cloning. For those who are pro-life, of course, the embryo and the fetus are already members of the human race, and it is wrong simply to destroy them. But even the federal directives for biological research, which do not admit the personhood of embryos, nonetheless demand that they be treated with "profound respect." And a law banning only reproductive cloning would produce, for the first time in federal statutes, a class of embryos it is a crime *not* to destroy, a class of embryos that must *not* be treated with profound respect.

Recent events in England are instructive. On April 19, health secretary Alan Milburn announced, to great fanfare in the British press, that Britain would shortly become the first country in the world to ban human cloning. But all he really meant was that Britain would prohibit reproduction by cloning, while continuing to promote the actual practice of cloning by encouraging laboratories to perfect their techniques. It was as polished an example of studied disingenuousness and blatant obfuscation as one will ever see. Four days later, the head of Britain's embryology authority quietly announced that scientists who had gone abroad to do embryo research illegal in Britain could return to "continuing acclaim."

For America, the lesson is clear: The only way to stop human reproductive cloning is to ban all human cloning, and to ban it now. There is no middle ground here, not merely because the principles involved do not admit it, but because the actual practice grants no room for compromise. To allow human cloning for medical and biological research is necessarily to allow—in the very near future—cloning for the reproduction of human beings.

TESTIMONY

U.S. Senate

William Kristol

President Bush quoted only one thinker in his unadorned—and quite effective—State of the Union address two months ago: Yogi Berra. The president commended to his congressional audience Mr. Berra's famous dictum, "When you come to a fork in the road, take it."

The president was preaching to the choir. American politicians don't like having to make difficult choices. Who can blame them? They have to balance diverse interests and juggle competing demands while doing justice to differing views among the citizens they represent. To govern is to choose, we're sometimes told. But, often, to govern in a big, pluralistic democracy like ours is *not* to choose, or not to choose too starkly, certainly not to choose irrevocably. After all, lots of choices are false choices; lots of bold decisions turn out badly. Avoiding forks in the road often isn't a bad idea.

George W. Bush knows this. After all, he's neither a conservative nor a moderate—he's a compassionate conservative. He wants to cut taxes—but also to increase government spending. He wants to cut regulations—but also to reassure environmentalists. He wants to strengthen our commitment to Taiwan—but also to work with Beijing. All of this is reasonable enough. And it's characteristic of politics in a Madisonian republic.

But a Madisonian republic has its Lincolnian moments. Occasionally, there really *is* a fork in the road. Occasionally, to govern *is* to choose—and not to choose is not to govern.

Selections from testimony before the U.S. Senate Committee on Commerce, Science, and Transportation, Subcommittee on Science, Technology, and Space, May 2, 2001.

Two generations ago, we had to choose whether to overcome segregation and discrimination. Under Lyndon Johnson's leadership, we made that choice. One generation ago, we had to choose whether we would try to overcome Communism abroad. Ronald Reagan led us in making that choice.

Today, we face a decision at least as momentous: whether we stumble heedlessly into a brave new world of eugenic enhancement and technological manufacture of human beings, or whether we will avert such a future. President Bush will lead us—or will fail to lead us—in that choice.

We are at an extraordinary moment of scientific progress, and scientific peril. The genetic revolution offers great hope for the medical treatment of disease, through gene therapy and other forms of healing. But if this revolution is not subject to human guidance and limitation, it will produce consequences that will be detrimental—no, devastating—to human liberty and human dignity.

These consequences have been laid out in detail, and the arguments against them made with great distinction, by thinkers ranging from Hans Jonas and Paul Ramsey a few decades ago to Leon Kass and Gilbert Meilaender today. But for current, practical purposes, our political leaders do not have to have studied all these arguments. All our politicians have to do now is to realize that, if they do not call a halt to certain experiments, if they do not limit the "progress" of science in certain ways, it will be virtually impossible to do so later . . .

But isn't it hopeless? Doesn't modernity mean that technology always trumps politics? Isn't scientific "progress" unstoppable?

No. No more than Communist domination of half the world was unstoppable, or that the further use of nuclear weapons after 1945 was unstoppable. No more than racial bigotry was unchangeable.

And in any case, to bow to the inevitability of this kind of scientific "progress" is to give up on the core of the American experiment: "that honorable determination," as Madison put it in *Federalist No. 39*, "to rest all our political experiments on the capacity of mankind for self-government." Science and technology may pose an even greater challenge to this determination than did slavery or communism. But to succumb is to forego our claim to self-government.

What, now, is to be done? The cloning of human beings is on the horizon. Ban it. . . . President Bush has spoken eloquently about his hope of ushering in a new "responsibility era." What greater responsibility do we have than halting a brave new world—one that, to quote Leon Kass, would put "human nature itself on the operating table, ready for alteration, 'enhancement,' and wholesale redesign"? A ban on human cloning would only be a first step down the road of responsibility and self-government—but it would be an important step.

PREVENTING A BRAVE NEW WORLD

Leon R. Kass

I.

The urgency of the great political struggles of the twentieth century, successfully waged against totalitarianisms first right and then left, seems to have blinded many people to a deeper and ultimately darker truth about the present age: all contemporary societies are travelling briskly in the same utopian direction. All are wedded to the modern technological project; all march eagerly to the drums of progress and fly proudly the banner of modern science; all sing loudly the Baconian anthem, "Conquer nature, relieve man's estate." Leading the triumphal procession is modern medicine, which is daily becoming ever more powerful in its battle against disease, decay, and death, thanks especially to astonishing achievements in biomedical science and technology—achievements for which we must surely be grateful.

Yet contemplating present and projected advances in genetic and reproductive technologies, in neuroscience and psychopharmacology, and in the development of artificial organs and computer-chip implants for human brains, we now clearly recognize new uses for biotechnical power that soar beyond the traditional medical goals of healing disease and relieving suffering. Human nature itself lies on the operating table, ready for alteration, for eugenic and psychic "enhancement," for wholesale re-design. In leading laboratories, academic and industrial, new creators are confidently amassing their powers and quietly honing their skills, while on the street their evangelists are

This essay originally appeared in the *New Republic* on May 21, 2001.

zealously prophesying a post-human future. For anyone who cares about preserving our humanity, the time has come to pay attention.

Some transforming powers are already here. The Pill. In vitro fertilization. Bottled embryos. Surrogate wombs. Cloning. Genetic screening. Genetic manipulation. Organ harvesting. Mechanical spare parts. Chimeras. Brain implants. Ritalin for the young, Viagra for the old, Prozac for everyone. And, to leave this vale of tears, a little extra morphine accompanied by Muzak.

Years ago Aldous Huxley saw it coming. In his charming but disturbing novel, *Brave New World* (it appeared in 1932 and is more powerful on each re-reading), he made its meaning strikingly visible for all to see. Unlike other frightening futuristic novels of the past century, such as Orwell's already dated *Nineteen Eighty-Four*, Huxley shows us a dystopia that goes with, rather than against, the human grain. Indeed, it is animated by our own most humane and progressive aspirations. Following those aspirations to their ultimate realization, Huxley enables us to recognize those less obvious but often more pernicious evils that are inextricably linked to the successful attainment of partial goods.

Huxley depicts human life seven centuries hence, living under the gentle hand of humanitarianism rendered fully competent by genetic manipulation, psychoactive drugs, hypnopaedia, and high-tech amusements. At long last, mankind has succeeded in eliminating disease, aggression, war, anxiety, suffering, guilt, envy, and grief. But this victory comes at the heavy price of homogenization, mediocrity, trivial pursuits, shallow attachments, debased tastes, spurious contentment, and souls without loves or longings. The Brave New World has achieved prosperity, community, stability, and nigh-universal contentment, only to be peopled by creatures of human shape but stunted humanity. They consume, fornicate, take "soma," enjoy "centrifugal bumble-puppy," and operate the machinery that makes it all possible. They do not read, write, think, love, or govern themselves. Art and science, virtue and religion, family and friendship are all passé. What matters most is bodily health and immediate gratification: "Never put off till tomorrow the fun you can have today." Brave New Man is so dehumanized that he does not even recognize what has been lost.

Huxley's novel, of course, is science fiction. Prozac is not yet Huxley's "soma"; cloning by nuclear transfer or splitting embryos is not exactly "Bokanovskification"; MTV and virtual-reality parlors are not quite the "feelies"; and our current safe and consequenceless sexual practices are not universally as loveless or as empty as those in the novel. But the kinships are disquieting, all the more so since our technologies of bio-psycho-engineering are still in their infancy, and in ways that make all too clear what they might look like in their full maturity. Moreover, the cultural changes that technology has already wrought among us should make us even more worried than Huxley would have us be.

In Huxley's novel, everything proceeds under the direction of an omnipotent—albeit benevolent—world state. Yet the dehumanization that he portrays does not really require despotism or external control. To the contrary, precisely because the society of the future will deliver exactly what we most want—health, safety, comfort, plenty, pleasure, peace of mind and length of days—we can reach the same humanly debased condition solely on the basis of free human choice. No need for World Controllers. Just give us the technological imperative, liberal democratic society, compassionate humanitarianism, moral pluralism, and free markets, and we can take ourselves to a Brave New World all by ourselves—and without even deliberately deciding to go. In case you had not noticed, the train has already left the station and is gathering speed, but no one seems to be in charge.

Some among us are delighted, of course, by this state of affairs: some scientists and biotechnologists, their entrepreneurial backers, and a cheering claque of sci-fi enthusiasts, futurologists, and libertarians. There are dreams to be realized, powers to be exercised, honors to be won, and money—big money—to be made. But many of us are worried, and not, as the proponents of the revolution self-servingly claim, because we are either ignorant of science or afraid of the unknown. To the contrary, we can see all too clearly where the train is headed, and we do not like the destination. We can distinguish cleverness about means from wisdom about ends, and we are loath to entrust the future of the race to those who cannot tell the difference. No friend of humanity cheers for a post-human future.

Yet for all our disquiet, we have until now done nothing to prevent it. We hide our heads in the sand because we enjoy the blessings that medicine keeps supplying, or we rationalize our inaction by declaring that human engineering is inevitable and we can do nothing about it. In either case, we are complicit in preparing for our own degradation, in some respects more to blame than the bio-zealots who, however misguided, are putting their money where their mouth is. Denial and despair, unattractive outlooks in any situation, become morally reprehensible when circumstances summon us to keep the world safe for human flourishing. Our immediate ancestors, taking up the challenge of their time, rose to the occasion and rescued the human future from the cruel dehumanizations of Nazi and Soviet tyranny. It is our more difficult task to find ways to preserve it from the soft dehumanizations of well-meaning but hubristic biotechnical "re-creationism"—and to do it without undermining biomedical science or rejecting its genuine contributions to human welfare.

Truth be told, it will not be easy for us to do so, and we know it. But rising to the challenge requires recognizing the difficulties. For there are indeed many features of modern life that will conspire to frustrate efforts aimed at the human control of the biomedical project. First, we Americans believe in technological automatism: where we do not foolishly believe that all innovation is progress, we fatalistically believe that it is inevitable ("If it can be done, it will be done, like it or not"). Second, we believe in freedom: the freedom of scientists to inquire, the freedom of technologists to develop, the freedom of entrepreneurs to invest and to profit, the freedom of private citizens to make use of existing technologies to satisfy any and all personal desires, including the desire to reproduce by whatever means. Third, the biomedical enterprise occupies the moral high ground of compassionate humanitarianism, upholding the supreme values of modern life—cure disease, prolong life, relieve suffering—in competition with which other moral goods rarely stand a chance. ("What the public wants is not to be sick," says James Watson, "and if we help them not to be sick, they'll be on our side.")

There are still other obstacles. Our cultural pluralism and easygoing relativism make it difficult to reach consensus on what we should embrace and what we should oppose; and moral objections to this or that biomedical practice are often facilely dismissed as religious or sectarian. Many people are unwilling to pronounce judgments about what is good or bad, right and wrong, even in matters of great importance, even for themselves—never mind for others or for society as a whole. It does not help that the biomedical project is now deeply entangled with commerce: there are increasingly powerful economic interests in favor of going full steam ahead, and no economic interests in favor of going slow. Since we live in a democracy, moreover, we face political difficulties in gaining a consensus to direct our future, and we have almost no political experience in trying to curtail the development of any new biomedical technology. Finally, and perhaps most troubling, our views of the meaning of our humanity have been so transformed by the scientific-technological approach to the world that we are in danger of forgetting what we have to lose, humanly speaking.

But though the difficulties are real, our situation is far from hopeless. Regarding each of the aforementioned impediments, there is another side to the story. Though we love our gadgets and believe in progress, we have lost our innocence regarding technology. The environmental movement especially has alerted us to the unintended damage caused by unregulated technological advance, and has taught us how certain dangerous practices can be curbed. Though we favor freedom of inquiry, we recognize that experiments are deeds and not speeches, and we prohibit experimentation on human subjects without their consent, even when cures from disease might be had by unfettered research; and we limit so-called reproductive freedom by proscribing incest, polygamy, and the buying and selling of babies.

Although we esteem medical progress, biomedical institutions have ethics committees that judge research proposals on moral grounds, and, when necessary, uphold the primacy of human freedom and human dignity even over scientific discovery. Our moral pluralism notwithstanding, national commissions and review bodies have sometimes reached moral consensus to recommend limits on permissible scientific research and technological application. On the economic front, the patenting of genes and life forms and the rapid rise of genomic commerce have elicited strong concerns and criticisms, leading even former enthusiasts of the new biology to recoil from the impending commodification of human life. Though we lack political institutions experienced in setting limits on biomedical innovation, federal agencies years ago rejected the development of the plutonium-powered artificial heart, and we have nationally prohibited commercial traffic in organs for transplantation, even though a market would increase the needed supply. In recent years, several American states and many foreign countries have successfully taken political action, making certain practices illegal and placing others under moratoriums (the creation of human embryos solely for research; human germ-line genetic alteration). Most importantly, the majority of Americans are not yet so degraded or so cynical as to fail to be revolted by the society depicted in Huxley's novel. Though the obstacles to effective action are significant, they offer no excuse for resignation. Besides, it would be disgraceful to concede defeat even before we enter the fray.

Not the least of our difficulties in trying to exercise control over where biology is taking us is the fact that we do not get to decide, once and for all, for or against the destination of a post-human world. The scientific discoveries and the technical powers that will take us there come to us piecemeal, one at a time and seemingly independent from one another, each often attractively introduced as a measure that will "help [us] not to be sick." But sometimes we come to a clear fork in the road where decision is possible, and where we know that our decision will make a world of difference—indeed, it will make a permanently different world. Fortunately, we stand now at the point of such a momentous decision. Events have conspired to provide us with a perfect opportunity to seize the initiative and to gain some control of the biotechnical project. I refer to the prospect of human cloning, a practice absolutely central to Huxley's fictional world. Indeed, creating and manipulating life in the laboratory is the gateway to a Brave New World, not only in fiction but also in fact.

"To clone or not to clone a human being" is no longer a fanciful question. Success in cloning sheep, and also cows, mice, pigs, and goats, makes it perfectly clear that a fateful decision is now at hand: whether we should welcome or even tolerate the cloning of human beings. If recent newspaper reports are to be believed, reputable scientists and physicians have announced their intention to produce the first human clone in the coming year. Their efforts may already be under way.

The media, gawking and titillating as is their wont, have been softening us up for this possibility by turning the bizarre into the familiar. In the four years since the birth of Dolly the cloned sheep, the tone of discussing the prospect of human cloning has gone from "Yuck" to "Oh?" to "Gee whiz" to "Why not?" The sentimentalizers, aided by leading bioethicists, have downplayed talk about eugenically cloning the beautiful and the brawny or the best and the brightest. They have taken instead to defending clonal reproduction for humanitarian or compassionate reasons: to treat infertility in people who are said to "have no other choice," to avoid the risk of severe genetic disease, to "replace" a child who has died. For the sake of these rare benefits, they would have us countenance the entire practice of human cloning, the consequences be damned.

But we dare not be complacent about what is at issue, for the stakes are very high. Human cloning, though partly continuous with previous reproductive technologies, is also something radically new in itself and in its easily foreseeable consequences—especially when coupled with powers for genetic "enhancement" and germline genetic modification that may soon become available, owing to the recently completed Human Genome Project. I exaggerate somewhat, but in the direction of the truth: we are compelled to decide nothing less than whether human procreation is going to remain human, whether children are going to be made to order rather than begotten, and whether we wish to say yes in principle to the road that leads to the dehumanized hell of *Brave New World*.

Four years ago I addressed this subject in these pages, trying to articulate the moral grounds of our repugnance at the prospect of human cloning ("The Wisdom of Repugnance," TNR, June 2, 1997). Subsequent events have only strengthened my conviction that cloning is a bad idea whose time should not come; but my emphasis this time is more practical. To be sure, I would still like to persuade undecided readers that cloning is a serious evil, but I am more interested in encouraging those who oppose human cloning but who think that we are impotent to prevent it, and in mobilizing them to support new and solid legislative efforts to stop it. In addition, I want readers who may worry less about cloning and more about the impending prospects of germline genetic manipulation or other eugenic practices to realize the unique practical opportunity that now presents itself to us.

For we have here a golden opportunity to exercise some control over where biology is taking us. The technology of cloning is discrete and well defined, and it requires considerable technical know-how and dexterity; we can therefore know by name many of the likely practitioners. The public demand for cloning is extremely low, and most people are decidedly against it. Nothing scientifically or medically important would be lost by banning clonal reproduction; alternative and non-objectionable means are available to obtain some of the most important medical benefits claimed for (non-reproductive) human cloning. The commercial interests in human cloning are, for now, quite limited; and the nations of the world are actively seeking to prevent it. Now may be as good a chance as we will ever have to get our hands on the wheel of the runaway train now headed for a post-human world and to steer it toward a more dignified human future.

II.

What is cloning? Cloning, or asexual reproduction, is the production of individuals who are genetically identical to an already existing individual. The procedure's name is fancy—"somatic cell nuclear transfer"—but its concept is simple. Take a mature but unfertilized egg; remove or deactivate its nucleus; introduce a nucleus obtained from a specialized (somatic) cell of an adult organism. Once the egg begins to divide, transfer the little embryo to a woman's uterus to initiate a pregnancy. Since almost all the hereditary material of a cell is contained within its nucleus, the re-nucleated egg and the individual into which it develops are genetically identical to the organism that was the source of the transferred nucleus.

An unlimited number of genetically identical individuals—the group, as well as each of its members, is called "a clone"—could be produced by nuclear transfer. In principle, any person, male or female, newborn or adult, could be cloned, and in any quantity; and because stored cells can outlive their sources, one may even clone the dead. Since cloning requires no personal involvement on the part of the person whose genetic material is used, it could easily be used to reproduce living or deceased persons without their consent—a threat to reproductive freedom that has received relatively little attention.

Some possible misconceptions need to be avoided. Cloning is not Xeroxing: the clone of Bill Clinton, though his genetic double, would enter the world hairless, toothless, and peeing in his diapers, like any other human infant. But neither is cloning just like natural twinning: the cloned twin will be identical to an older, existing adult; and it will arise not by chance but by deliberate design; and its entire genetic makeup will be pre-selected by its parents and/or scientists. Moreover, the success rate of cloning, at least at first, will probably not be very high: the Scots transferred two hundred seventy-seven adult nuclei into sheep eggs, implanted twenty-nine clonal embryos, and achieved the birth of only one live lamb clone.

For this reason, among others, it is unlikely that, at least for now, the practice would be very popular; and there is little immediate worry of mass-scale production of multicopies. Still, for the tens of thousands of people who sustain more than three hundred assisted-reproduction clinics in the United States and already avail themselves of in vitro fertilization and other techniques, cloning would be an option with virtually no added fuss. Panos Zavos, the Kentucky reproduction specialist who has announced his plans to clone a child, claims that he has already received thousands of e-mailed requests from people eager to clone, despite the known risks of failure and damaged offspring. Should commercial interests develop in "nucleus-banking," as they have in sperm-banking and egg-harvesting; should famous athletes or other celebrities decide to market their DNA the way they now market their autographs and nearly everything else; should techniques of embryo and germline genetic testing and manipulation arrive as anticipated, increasing the use of laboratory assistance in order to obtain "better" babies—should all this come to pass, cloning, if it is permitted, could become more than a marginal practice simply on the basis of free reproductive choice.

What are we to think about this prospect? Nothing good. Indeed, most people are repelled by nearly all aspects of human cloning: the possibility of mass production of human beings, with large clones of look-alikes, compromised in their individuality; the idea of father-son or mother-daughter "twins"; the bizarre prospect of a woman bearing and rearing a genetic copy of herself, her spouse, or even her deceased father or mother; the grotesqueness of conceiving a child as an exact "replacement" for another who has died; the utilitarian creation of embryonic duplicates of oneself, to be frozen away or created when needed to provide homologous tissues or organs for transplantation; the narcissism of those who would clone themselves, and the arrogance of others who think they know who deserves to be cloned; the Frankensteinian hubris to create a human life and increasingly to control its destiny; men playing at being God. Almost no one finds any of the suggested reasons for human cloning compelling, and almost everyone anticipates its possible misuses and abuses. And the popular belief that human cloning cannot be prevented makes the prospect all the more revolting.

Revulsion is not an argument; and some of yesterday's repugnances are today calmly accepted—not always for the better. In some crucial cases, however, repugnance is the emotional expression of deep wisdom, beyond reason's power completely to articulate it. Can anyone really give an argument fully adequate to the horror that is father-daughter incest (even with consent), or bestiality, or the mutilation of a corpse, or the eating of human flesh, or the rape or murder of another human being? Would anybody's failure to give full rational justification for his revulsion at those practices make that revulsion ethically suspect?

I suggest that our repugnance at human cloning belongs in this category. We are repelled by the prospect of cloning human beings not because of the strangeness or the novelty of the undertaking, but because we intuit and we feel, immediately and without argument, the violation of things that we rightfully hold dear. We sense that cloning represents a profound defilement of our given nature as procreative beings, and of the social relations built on this natural ground. We also sense that cloning is a radical form of child abuse. In this age in which everything is held to be permissible so long as it is freely done, and in which our bodies are regarded as mere instruments of our autonomous rational will, repugnance may be the only voice left that speaks up to defend the central core of our humanity. Shallow are the souls that have forgotten how to shudder.

III.

Yet repugnance need not stand naked before the bar of reason. The wisdom of our horror at human cloning can be at least partially articulated, even if this is finally one of those instances about which the heart has its reasons that reason cannot entirely know. I offer four objections to human cloning: that it constitutes unethical experimentation; that it threatens identity and individuality; that it turns procreation into manufacture (especially when understood as the harbinger of manipulations to come); and that it means despotism over children and perversion of parenthood. Please note: I speak only about so-called reproductive cloning, not about the creation of cloned embryos for research. The objections that may be raised against creating (or using) embryos for research are entirely independent of whether the research embryos are produced by cloning. What is radically distinct and radically new is reproductive cloning.

Any attempt to clone a human being would constitute an unethical experiment upon the resulting child-to-be. In all the animal experiments, fewer than two to three percent of all cloning attempts succeeded. Not only are there fetal deaths and stillborn infants, but many of the so-called "successes" are in fact failures. As has only recently become clear, there is a very high incidence of major disabilities and deformities in cloned animals that attain live birth. Cloned cows often have heart and lung problems; cloned mice later develop pathological obesity; other live-born cloned animals fail to reach normal developmental milestones.

The problem, scientists suggest, may lie in the fact that an egg with a new somatic nucleus must re-program itself in a matter of minutes or hours (whereas the nucleus of an unaltered egg has been prepared over months and years). There is thus a greatly increased likelihood of error in translating the genetic instructions, leading to developmental defects some of which will show themselves only much later. (Note also that these induced abnormalities may also affect the stem cells that scientists hope to harvest from cloned embryos. Lousy embryos, lousy stem cells.) Nearly all scientists now agree that attempts to clone human beings carry massive risks of producing unhealthy, abnormal, and malformed children. What are we to do with them? Shall we just discard the ones that fall short of expectations? Considered opinion is today nearly unanimous, even among scientists: attempts at human cloning are irresponsible and unethical. We cannot ethically even get to know whether or not human cloning is feasible.

If it were successful, cloning would create serious issues of identity and individuality. The clone may experience concerns about his distinctive identity not only because he will be, in genotype and in appearance, identical to another human being, but because he may also be twin to the person who is his "father" or his "mother"—if one can still call them that. Unaccountably, people treat as innocent the homey case of intra-familial cloning—the cloning of husband or wife (or single mother). They forget about the unique dangers of mixing the twin relation with the parent-child relation. (For this situation, the relation of contemporaneous twins is no precedent; yet even this less problematic situation teaches us how difficult it is to wrest independence from the being for whom one has the most powerful affinity.) Virtually no parent is going to be able to treat a clone of himself or herself as one treats a child generated by the lottery of sex. What will happen when the adolescent clone of Mommy becomes the spitting image of the woman with whom Daddy once fell in love? In case of divorce, will Mommy still love the clone of Daddy, even though she can no longer stand the sight of Daddy himself?

Most people think about cloning from the point of view of adults choosing to clone. Almost nobody thinks about what it would be like to be the cloned child. Surely his or her new life would constantly be scrutinized in relation to that of the older version. Even in the absence of unusual parental expectations for the clone—say, to live the same life, only without its errors—the child is likely to be ever a curiosity, ever a potential source of déjà vu. Unlike "normal" identical twins, a cloned individual—copied from whomever—will be saddled with a genotype that has already lived. He will not be fully a surprise to the world: people are likely always to compare his doings in life with those of his alter ego, especially if he is a clone of someone gifted or famous. True, his nurture and his circumstance will be different; genotype is not exactly destiny. But one must also expect parental efforts to shape this new life after the original—or at least to view the child with the original version always firmly in mind. For why else did they clone from the star basketball player, the mathematician, or the beauty queen—or even dear old Dad—in the first place?

Human cloning would also represent a giant step toward the transformation of begetting into making, of procreation into manufacture (literally, "handmade"), a process that has already begun with in vitro fertilization and genetic testing of embryos. With cloning, not only is the process in hand, but the total genetic blueprint of the cloned individual is selected and determined by the human artisans. To be sure, subsequent development is still according to natural processes; and the resulting children will be recognizably human. But we would be taking a major step into making man himself simply another one of the man-made things.

How does begetting differ from making? In natural procreation, human beings come together to give existence to another being that is formed exactly as we were, by what we are—living, hence perishable, hence aspiringly erotic, hence procreative human beings. But in clonal reproduction, and in the more advanced forms of manufacture to which it will lead, we give existence to a being not by what we are but by what we intend and design.

Let me be clear. The problem is not the mere intervention of technique, and the point is not that "nature knows best." The problem is that any child whose being, character, and capacities exist owing to human design does not stand on the same plane as its makers. As with any product of our making, no matter how excellent, the artificer stands above it, not as an equal but as a superior, transcending it by his will and creative prowess. In human cloning, scientists and prospective "parents" adopt a technocratic attitude toward human children: human children become their artifacts. Such an arrangement is profoundly dehumanizing, no matter how good the product.

Procreation dehumanized into manufacture is further degraded by commodification, a virtually inescapable result of allowing baby-making to proceed under the banner of commerce. Genetic and reproductive biotechnology companies are already growth industries, but they will soon go into commercial orbit now that the Human Genome Project has been completed. "Human eggs for sale" is already a big business, masquerading under the pretense of "donation." Newspaper advertisements on elite college campuses offer up to $50,000 for an egg "donor" tall enough to play women's basketball and with SAT scores high enough for admission to Stanford; and to nobody's surprise, at such prices there are many young coeds eager to help shoppers obtain the finest babies money can buy. (The egg and womb-renting entrepreneurs shamelessly proceed on the ancient, disgusting, misogynist premise that most women will give you access to their bodies, if the price is right.) Even before the capacity for human cloning is perfected, established companies will have invested in the harvesting of eggs from ovaries obtained at autopsy or through ovarian surgery, practiced embryonic genetic alteration, and initiated the stockpiling of prospective donor tissues. Through the rental of surrogate-womb services, and through the buying and selling of tissues and embryos priced according to the merit of the donor, the commodification of nascent human life will be unstoppable.

Finally, the practice of human cloning by nuclear transfer—like other anticipated forms of genetically engineering the next generation—would enshrine and aggravate a profound misunderstanding of the meaning of having children and of the parent-child relationship. When a couple normally chooses to procreate, the partners are saying yes to the emergence of new life in its novelty—are saying yes not only to having a child, but also to having whatever child this child turns out to be. In accepting our finitude, in opening ourselves to our replacement, we tacitly confess the limits of our control.

Embracing the future by procreating means precisely that we are relinquishing our grip in the very activity of taking up our own share in what we hope will be the immortality of human life and the human species. This means that our children are not our children: they are not our property, they are not our possessions. Neither are they supposed to live our lives for us, or to live anyone's life but their own. Their genetic distinctiveness and independence are the natural foreshadowing of the deep truth that they have their own, never-before-enacted life to live. Though sprung from a past, they take an uncharted course into the future.

Much mischief is already done by parents who try to live vicariously through their children. Children are sometimes compelled to fulfill the broken dreams of unhappy parents. But whereas most parents normally have hopes for their children, cloning parents will have expectations. In cloning, such overbearing parents will have taken at the start a decisive step that contradicts the entire meaning of the open and forward-looking nature of parent-child relations. The child is given a genotype that has already lived, with full expectation that this blueprint of a past life ought to be controlling the life that is to come. A wanted child now means a child who exists precisely to fulfill parental wants. Like all the more precise eugenic manipulations that will follow in its wake, cloning is thus inherently despotic, for it seeks to make one's children after one's own image (or an image of one's choosing) and their future according to one's will.

Is this hyperbolic? Consider concretely the new realities of responsibility and guilt in the households of the cloned. No longer only the sins of the parents, but also the genetic choices of the parents, will be visited on the children—and beyond the third and fourth generation; and everyone will know who is responsible. No parent will be able to blame nature or the lottery of sex for an unhappy adolescent's big nose, dull wit, musical ineptitude, nervous disposition, or anything else that he hates about himself. Fairly or not, children will hold their cloners responsible for everything, for nature as well as for nurture. And parents, especially the better ones, will be limitlessly liable to guilt. Only the truly despotic souls will sleep the sleep of the innocent.

IV.

The defenders of cloning are not wittingly friends of despotism. Quite the contrary. Deaf to most other considerations, they regard themselves mainly as friends of freedom: the freedom of individuals to reproduce, the freedom of scientists and inventors to discover and to devise and to foster "progress" in genetic knowledge and technique, the freedom of entrepreneurs to profit in the market. They want large-scale cloning only for animals, but they wish to preserve cloning as a human option for exercising our "right to reproduce"—our right to have children, and children with "desirable genes." As some point out, under our "right to reproduce" we already practice early forms of unnatural, artificial, and extra-marital reproduction, and we already practice early forms of eugenic choice. For that reason, they argue, cloning is no big deal.

We have here a perfect example of the logic of the slippery slope. The principle of reproductive freedom currently enunciated by the proponents of cloning logically embraces the ethical acceptability of sliding all the way down: to producing children wholly in the laboratory from sperm to term (should it become feasible), and to producing children whose entire genetic makeup will be the product of parental eugenic planning and choice. If reproductive freedom means the right to have a child of one's own choosing by whatever means, then reproductive freedom knows and accepts no limits.

Proponents want us to believe that there are legitimate uses of cloning that can be distinguished from illegitimate uses, but by their own principles no such limits can be found. (Nor could any such limits be enforced in practice: once cloning is permitted, no one ever need discover whom one is cloning and why.) Reproductive freedom, as they understand it, is governed solely by the subjective wishes of the parents-to-be. The sentimentally appealing case of the childless married couple is, on these grounds, indistinguishable from the case of an individual (married or not) who would like to clone someone famous or talented, living or dead. And the principle here endorsed justifies not only cloning but also all future artificial attempts to create (manufacture) "better" or "perfect" babies.

The "perfect baby," of course, is the project not of the infertility doctors, but of the eugenic scientists and their supporters, who, for the time being, are content to hide behind the skirts of the partisans of reproductive freedom and compassion for the infertile. For them, the paramount right is not the so-called right to reproduce, it is what the biologist Bentley Glass called, a quarter of a century ago, "the right of every child to be born with a sound physical and mental constitution, based on a sound genotype . . . the inalienable right to a sound heritage." But to secure this right, and to achieve the requisite quality control over new human life, human conception and gestation will need to be brought fully into the bright light of the laboratory, beneath which the child-to-be can be fertilized, nourished, pruned, weeded, watched, inspected, prodded, pinched, cajoled, injected, tested, rated, graded, approved, stamped, wrapped, sealed, and delivered. There is no other way to produce the perfect baby.

If you think that such scenarios require outside coercion or governmental tyranny, you are mistaken. Once it becomes possible, with the aid of human genomics, to produce or to select for what some regard as "better babies"—smarter, prettier, healthier, more athletic—parents will leap at the opportunity to "improve" their offspring. Indeed, not to do so will be socially regarded as a form of child neglect. Those who would ordinarily be opposed to such tinkering will be under enormous pressure to compete on behalf of their as yet unborn children—just as some now plan almost from their children's birth how to get them into Harvard. Never mind that, lacking a standard of "good" or "better," no one can really know whether any such changes will truly be improvements.

Proponents of cloning urge us to forget about the science-fiction scenarios of laboratory manufacture or multiple-copy clones, and to focus only on the sympathetic cases of infertile couples exercising their reproductive rights. But why, if the single cases are so innocent, should multiplying their performance be so off-putting? (Similarly, why do others object to people's making money from that practice if the practice itself is perfectly acceptable?) The so-called science-fiction cases—say, *Brave New World*—make vivid the meaning of what looks to us, mistakenly, to be benign. They reveal that what looks like compassionate humanitarianism is, in the end, crushing dehumanization.

V.

Whether or not they share my reasons, most people, I think, share my conclusion: that human cloning is unethical in itself and dangerous in its likely consequences, which include the precedent that it will establish for designing our children. Some reach this conclusion for their own good reasons, different from my own: concerns about distributive justice in access to eugenic cloning; worries about the genetic effects of asexual "inbreeding"; aversion to the implicit premise of genetic determinism; objections to the embryonic and fetal wastage that must necessarily accompany the efforts; religious opposition to "man playing God." But never mind why: the overwhelming majority of our fellow Americans remain firmly opposed to cloning human beings.

For us, then, the real questions are: What should we do about it? How can we best succeed? These questions should concern everyone eager to secure deliberate human control over the powers that could re-design our humanity, even if cloning is not the issue over which they would choose to make their stand. And the answer to the first question seems pretty plain. What we should do is work to prevent human cloning by making it illegal.

We should aim for a global legal ban, if possible, and for a unilateral national ban at a minimum—and soon, before the fact is upon us. To be sure, legal bans can be violated; but we certainly curtail much mischief by outlawing incest, voluntary servitude, and the buying and selling of organs and babies. To be sure, renegade scientists may secretly undertake to violate such a law, but we can deter them by both criminal sanctions and monetary penalties, as well as by removing any incentive they have to proudly claim credit for their technological bravado.

Such a ban on clonal baby-making will not harm the progress of basic genetic science and technology. On the contrary, it will reassure the public that scientists are happy to proceed without violating the deep ethical norms and intuitions of the human community. It will also protect honorable scientists from a public backlash against the brazen misconduct of the rogues. As many scientists have publicly confessed, free and worthy science probably has much more to fear from a strong public reaction to a cloning fiasco than it does from a cloning ban, provided that the ban is judiciously crafted and vigorously enforced against those who would violate it.

Five states—Michigan, Louisiana, California, Rhode Island, and Virginia—have already enacted a ban on human cloning, and several others are likely to follow suit this year. Michigan, for example, has made it a felony, punishable by imprisonment for not more than ten years or a fine of not more than $10 million, or both, to "intentionally engage in or attempt to engage in human cloning," where human cloning means "the use of human somatic cell nuclear transfer technology to produce a human embryo." Internationally, the movement to ban human cloning gains momentum. France and Germany have banned cloning (and germline genetic engineering), and the Council of Europe is working to have it banned in all of its forty-one member countries, and Canada is expected to follow suit. The United Nations, UNESCO, and the Group of Seven have called for a global ban on human cloning.

Given the decisive actions of the rest of the industrialized world, the United States looks to some observers to be a rogue nation. A few years ago, soon after the birth of Dolly, President Clinton called for legislation to outlaw human cloning, and attempts were made to produce a national ban. Yet none was enacted, despite general agreement in Congress that it would be desirable to have such a ban. One might have thought that it would be easy enough to find clear statutory language for prohibiting attempts to clone a human being (and other nations have apparently not found it difficult). But, alas, in the last national go-around, there was trouble over the apparently vague term "human being," and whether it includes the early (pre-implantation) embryonic stages of human life. Learning from this past failure, we can do better this time around. Besides, circumstances have changed greatly in the intervening three years, making a ban both more urgent and less problematic.

Two major anti-cloning bills were introduced into the Senate in 1998. The Democratic bill (Kennedy-Feinstein) would have banned so-called reproductive cloning by prohibiting transfer of cloned embryos into women to initiate pregnancy. The Republican bill (Frist-Bond) would have banned *all* cloning by prohibiting the creation even of embryonic human clones. Both sides opposed "reproductive cloning," the attempt to bring to birth a living human child who is the clone of someone now (or previously) alive. But the Democratic bill sanctioned creating cloned embryos for research purposes, and the Republican bill did not. The pro-life movement could not support the former, whereas the scientific community and the biotechnology industry opposed the latter; indeed, they successfully lobbied a dozen Republican senators to oppose taking a vote on the Republican bill (which even its supporters now admit was badly drafted). Owing to a deep and unbridgeable gulf over the question of embryo research, we did not get the congressional ban on reproductive cloning that nearly everyone wanted. It would be tragic if we again failed to produce a ban on human cloning because of its seemingly unavoidable entanglement with the more divisive issue of embryo research.

To find a way around this impasse, several people (myself included) advocated a legislative "third way," one that firmly banned only reproductive cloning but did not legitimate creating cloned embryos for research. This, it turns out, is hard to do. It is easy enough to state the necessary negative disclaimer that would set aside the embryo-research question: "Nothing in this act shall be taken to determine the legality of creating cloned embryos for research; this act neither permits nor prohibits such activity." It is much more difficult to state the positive prohibition in terms that are unambiguous and acceptable to all sides. To indicate only one difficulty: indifference to the creation of embryonic clones coupled with a ban (only) on their transfer would place the federal government in the position of demanding the destruction of nascent life, a bitter pill to swallow even for pro-choice advocates.

Given both these difficulties, and given the imminence of attempts at human cloning, I now believe that what we need is an all-out ban on human cloning, including the creation of embryonic clones. I am convinced that all halfway measures will prove to be morally, legally, and strategically flawed, and—most important—that they will not be effective in obtaining the desired result. Anyone truly serious about preventing human reproductive cloning must seek to stop the process from the beginning. Our changed circumstances, and the now evident defects of the less restrictive alternatives, make an all-out ban by far the most attractive and effective option.

Here's why. Creating cloned human children ("reproductive cloning") necessarily begins by producing cloned human embryos. Preventing the latter would prevent the former, and prudence alone might counsel building such a "fence around the law." Yet some scientists favor embryo cloning as a way of obtaining embryos for research or as sources of cells and tissues for the possible benefit of others. (This practice they misleadingly call "therapeutic cloning" rather than the more accurate "cloning for research" or "experimental cloning," so as to obscure the fact that the clone will be "treated" only to exploitation and destruction, and that any potential future beneficiaries and any future "therapies" are at this point purely hypothetical.)

The prospect of creating new human life solely to be exploited in this way has been condemned on moral grounds by many people—including *The Washington Post*, President Clinton, and many other supporters of a woman's right to abortion—as displaying a profound disrespect for life. Even those who are willing to scavenge so-called "spare embryos"—those products of in vitro fertilization made in excess of people's reproductive needs, and otherwise likely to be discarded—draw back from creating human embryos explicitly and solely for research purposes. They reject outright what they regard as the exploitation and the instrumentalization of nascent human life. In addition, others who are agnostic about the moral status of the embryo see the wisdom of not needlessly offending the sensibilities of their fellow citizens who are opposed to such practices.

But even setting aside these obvious moral first impressions, a few moments of reflection show why an anti-cloning law that permitted the cloning of embryos but criminalized their transfer to produce a child would be a moral blunder. This would be a law that was not merely permissively "pro-choice" but emphatically and prescriptively "anti-life." While permitting the creation of an embryonic life, it would make it a federal offense to try to keep it alive and bring it to birth. Whatever one thinks of the moral status or the ontological status of the human embryo, moral sense and practical wisdom recoil from having the government of the United States on record as requiring the destruction of nascent life and, what is worse, demanding the punishment of those who would act to preserve it by (feloniously!) giving it birth.

But the problem with the approach that targets only reproductive cloning (that is, the transfer of the embryo to a woman's uterus) is not only moral but also legal and strategic. A ban only on reproductive cloning would turn out to be unenforceable. Once cloned embryos were produced and available in laboratories and assisted-reproduction centers, it would be virtually impossible to control what was done with them. Biotechnical experiments take place in laboratories, hidden from public view, and, given the rise of high-stakes commerce in biotechnology, these experiments are concealed from the competition. Huge stockpiles of cloned human embryos could thus be produced and bought and sold without anyone knowing it. As we have seen with in vitro embryos created to treat infertility, embryos produced for one reason can be used for another reason: today "spare embryos" once created to begin a pregnancy are now used in research, and tomorrow clones created for research will be used to begin a pregnancy.

Assisted reproduction takes place within the privacy of the doctor-patient relationship, making outside scrutiny extremely difficult. Many infertility experts probably would obey the law, but others could and would defy it with impunity, their doings covered by the veil of secrecy that is the principle of medical confidentiality. Moreover, the transfer of embryos to begin a pregnancy is a simple procedure (especially compared with manufacturing the embryo in the first place), simple enough that its final steps could be self-administered by the woman, who would thus absolve the doctor of blame for having "caused" the illegal transfer. (I have in mind something analogous to Kevorkian's suicide machine, which was designed to enable the patient to push the plunger and the good "doctor" to evade criminal liability.)

Even should the deed become known, governmental attempts to enforce the reproductive ban would run into a swarm of moral and legal challenges, both to efforts aimed at preventing transfer to a woman and—even worse—to efforts seeking to prevent birth after transfer has occurred. A woman who wished to receive the embryo clone would no doubt seek a judicial restraining order, suing to have the law overturned in the name of a constitutionally protected interest in her own reproductive choice to clone. (The cloned child would be born before the legal proceedings were complete.) And should an "illicit clonal pregnancy" be discovered, no governmental agency would compel a woman to abort the clone, and there would be an understandable storm of protest should she be fined or jailed after she gives birth. Once the baby is born, there would even be sentimental opposition to punishing the doctor for violating the law—unless, of course, the clone turned out to be severely abnormal.

For all these reasons, the only practically effective and legally sound approach is to block human cloning at the start, at the production of the embryo clone. Such a ban can be rightly characterized not as interference with reproductive freedom, nor even as interference with scientific inquiry, but as an attempt to prevent the unhealthy, unsavory, and unwelcome manufacture of and traffic in human clones.

VI.

Some scientists, pharmaceutical companies, and bio-entrepreneurs may balk at such a comprehensive restriction. They want to get their hands on those embryos, especially for their stem cells, those pluripotent cells that can in principle be turned into any cells and any tissues in the body, potentially useful for transplantation to repair somatic damage. Embryonic stem cells need not come from cloned embryos, of course; but the scientists say that stem cells obtained from clones could be therapeutically injected into the embryo's adult "twin" without any risk of immunological rejection. It is the promise of rejection-free tissues for transplantation that so far has been the most successful argument in favor of experimental cloning. Yet new discoveries have shown that we can probably obtain the same benefits without embryo cloning. The facts are much different than they were three years ago, and the weight in the debate about cloning for research should shift to reflect the facts.

Numerous recent studies have shown that it is possible to obtain highly potent stem cells from the bodies of children and adults—from the blood, bone marrow, brain, pancreas, and, most recently, fat. Beyond all expectations, these non-embryonic stem cells have been shown to have the capacity to turn into a wide variety of specialized cells and tissues. (At the same time, early human therapeutic efforts with stem cells derived from embryos have produced some horrible results, the cells going wild in their new hosts and producing other tissues in addition to those in need of replacement. If an in vitro embryo is undetectably abnormal—as so often they are—the cells derived from it may also be abnormal.) Since cells derived from our own bodies are more easily and cheaply available than cells harvested from specially manufactured clones, we will almost surely be able to obtain from ourselves any needed homologous transplantable cells and tissues, without the need for egg donors or cloned embryonic copies of ourselves. By pouring our resources into *adult* stem cell research (or, more accurately, "non-embryonic" stem cell research), we can also avoid the morally and legally vexing issues in embryo research. And more to our present subject, by eschewing the cloning of embryos, we make the cloning of human beings much less likely.

A few weeks ago an excellent federal anti-cloning bill was introduced in Congress, sponsored by Senator Sam Brownback and Representative David Weldon. This carefully drafted legislation seeks to prevent the cloning of human beings at the very first step, by prohibiting somatic cell nuclear transfer to produce embryonic clones, and provides substantial criminal and monetary penalties for violating the law. The bill makes very clear that there is to be no interference with the scientific and medically useful practices of cloning DNA fragments (molecular cloning), with the duplication of somatic cells (or stem cells) in tissue culture (cell cloning), or with whole-organism or embryo cloning of non-human animals. If enacted, this law would bring the United States into line with the current or soon-to-be-enacted practices of many other nations. Most important, it offers us the best chance—the only realistic chance—that we have to keep human cloning from happening, or from happening much.

Getting this bill passed will not be easy. The pharmaceutical and biotech companies and some scientific and patient-advocacy associations may claim that the bill is the work of bio-Luddites: anti-science, a threat to free inquiry, an obstacle to obtaining urgently needed therapies for disease. Some feminists and pro-choice groups will claim that this legislation is really only a sneaky device for fighting *Roe* v. *Wade*, and they will resist anything that might be taken even to hint that a human embryo has any moral worth. On the other side, some right-to-life purists, who care not how babies are made as long as life will not be destroyed, will withhold their support because the bill does not take a position against embryo twinning or embryo research in general.

All of these arguments are wrong, and all of them must be resisted. This is not an issue of pro-life versus pro-choice. It is not about death and destruction, or about a woman's right to choose. It is only and emphatically about baby design and manufacture: the opening skirmish of a long battle against eugenics and against a post-human future. As such, it is an issue that should not divide "the left" and "the right"; and there are people across the political spectrum who are coalescing in the efforts to stop human cloning. (The prime sponsor of Michigan's comprehensive anti-cloning law is a pro-choice Democratic legislator.) Everyone needs to understand that, whatever we may think about the moral status of embryos, once embryonic clones are produced in the laboratories the eugenic revolution will have begun. And we shall have lost our best chance to do anything about it.

As we argue in the coming weeks about this legislation, let us be clear about the urgency of our situation and the meaning of our action or inaction. Scientists and doctors whose names we know, and probably many others whose names we do not know, are today working to clone human beings. They are aware of the immediate hazards, but they are undeterred. They are prepared to screen and to destroy anything that looks abnormal. They do not care that they will not be able to detect most of the possible defects. So confident are they in their rectitude that they are willing to ignore all future consequences of the power to clone human beings. They are prepared to gamble with the well-being of any live-born clones, and, if I am right, with a great deal more, all for the glory of being the first to replicate a human being. They are, in short, daring the community to defy them. In these circumstances, our silence can only mean acquiescence. To do nothing now is to accept the responsibility for the deed and for all that follows predictably in its wake.

I appreciate that a federal legislative ban on human cloning is without American precedent, at least in matters technological. Perhaps such a ban will prove ineffective; perhaps it will eventually be shown to have been a mistake. (If so, it could later be reversed.) If enacted, however, it will have achieved one overwhelmingly important result, in addition to its contribution to thwarting cloning: it will place the burden of practical proof where it belongs. It will require the proponents to show very clearly what great social or medical good can be had only by the cloning of human beings. Surely it is only for such a compelling case, yet to be made or even imagined, that we should wish to risk this major departure—or any other major departure—in human procreation.

Americans have lived by and prospered under a rosy optimism about scientific and technological progress. The technological imperative has probably served us well, though we should admit that there is no accurate method for weighing benefits and harms. And even when we recognize the unwelcome outcomes of technological advance, we remain confident in our ability to fix all the "bad" consequences—by regulation or by means of still newer and better technologies. Yet there is very good reason for shifting the American paradigm, at least regarding those technological interventions into the human body and mind that would surely effect fundamental (and likely irreversible) changes in human nature, basic human relationships, and what it means to be a human being. Here we should not be willing to risk everything in the naive hope that, should things go wrong, we can later set them right again.

Some have argued that cloning is almost certainly going to remain a marginal practice, and that we should therefore permit people to practice it. Such a view is shortsighted. Even if cloning is rarely undertaken, a society in which it is tolerated is no longer the same society—any more than is a society that permits (even small-scale) incest or cannibalism or slavery. A society that allows cloning, whether it knows it or not, has tacitly assented to the conversion of procreation into manufacture and to the treatment of children as purely the projects of our will. Willy-nilly, it has acquiesced in the eugenic re-design of future generations. The humanitarian superhighway to a Brave New World lies open before this society.

But the present danger posed by human cloning is, paradoxically, also a golden opportunity. In a truly unprecedented way, we can strike a blow for the human control of the technological project, for wisdom, for prudence, for human dignity. The prospect of human cloning, so repulsive to contemplate, is the occasion for deciding whether we shall be slaves of unregulated innovation, and ultimately its artifacts, or whether we shall remain free human beings who guide our powers toward the enhancement of human dignity. The humanity of the human future is now in our hands.

SEPARATING GOOD BIOTECH FROM BAD

Francis Fukuyama

The Bush administration will soon be facing a decision on whether to continue the existing ban on federal funding for stem-cell research, or to accept guidelines established by the National Institutes of Health that would permit such research, while seeking to minimize the need for embryos derived from abortions.

Sources within the administration, as well as a number of conservatives, have argued for keeping the ban, on the ground that stem-cell research can proceed using adult stem cells. This position is misguided and sidesteps the serious long-term issue raised by stem cells, which is the need for an institutional framework for the regulation of all human biotechnology, and not just research funded by the federal government. The administration ought to permit federal funding of stem-cell research, but extend the guidelines to all research done in the U.S. . . .

There are two types of concerns motivating the current ban on federal funding for research on embryonic stem cells. The first arises out of a fear that the cells will come from aborted fetuses, and thereby either encourage abortion or else retroactively justify the decision to have an abortion. The second has to do with the very fact that embryos are destroyed when stem cells are derived from them, which is itself seen as an affront to their dignity as human beings.

The guidelines that the NIH has suggested for embryonic stem-cell research more than adequately answer the first concern over abortion. They forbid the use of aborted fetuses for such research, forbid the deliberate cre-

This essay originally appeared in the *Wall Street Journal* on May 23, 2001.

ation of an embryo for research purposes, and seek to derive stem cells from the extra embryos produced by in-vitro fertilization clinics that were destined for destruction or indefinite storage.

The fact is that scientists will not need a lot of embryos to proceed with this research, since a stem-cell "line," once isolated, can be propagated almost indefinitely. One such line has already been isolated by University of Wisconsin biologist James A. Thomson, working with private funding. The likelihood that future research will create a demand for aborted fetuses or encourage any more abortions than currently take place is very small.

The second objection is harder to answer if you believe that destroying a 32-cell blastocyst is morally equivalent to infanticide. (I personally do not, but take seriously the concerns of those who do.)

But those who accept that premise must ask themselves why it is also legal for an in-vitro fertilization clinic to destroy these embryos, but impermissible to use them for research. Or to put it another way, if their concern is the deliberate destruction of embryos, why are they not bending every effort to ban in-vitro fertilization altogether rather than focusing on stem cells, since the former leads to massively greater levels of harm to embryos? Could it be that they are motivated by a utilitarian concern that voters would never abide banning in-vitro fertilization? And if so, why not concede the utilitarian interest we all have in future stem-cell research?

It is perhaps to avoid having to face such questions that advocates of the ban have sought refuge in the argument that the use of embryonic stem cells is not necessary, and that whatever benefits stem cells will provide can come from adult cells, or the stem cells in umbilical cord blood. To date, adult stem cells have indeed produced some dramatic results. Researchers have used stem cells to successfully treat mice specially bred to have strokes, and to grow a kind of white blood cell that is resistant to the HIV virus.

By contrast, there have been recent setbacks in therapies using embryonic stem cells. The *New England Journal of Medicine* has reported on an experiment in which fetal-cell transplants into the brains of patients during a clinical trial yielded extreme and irreversible side effects such as uncontrollable writhing and jerking. Pluripotent stem cells may in fact be too powerful, and, like the cancer cells to which they bear a certain resemblance, difficult to control.

Nevertheless, the argument that adult stem cells will provide all the benefits expected of embryonic ones is a dodge. It may be that they are right, but neither they nor anyone else will know this for certain until the research is done.

As Charles Krauthammer has pointed out, American conservatives, with their single-minded focus on abortion, are missing the boat on what is really at stake in contemporary biotechnology by worrying about the source of stem cells rather than their ultimate destiny. We should instead be asking whether we really want to manufacture new human body parts in the chest cavities of pigs, or whether immortality is in fact the appropriate goal of stem-cell research, as some of its proponents suggest.

Another issue that Congress has thus far failed to address is the need for a comprehensive regulatory framework for human biotechnology. Such rules as exist regarding issues like stem cells, germ-line engineering, human cloning, and human experimentation more broadly have focused on federally funded research.

This was fine in an age when the NIH funded the vast majority of biotech research. But today, there is a huge private biotech industry and hundreds of millions of loose research dollars seeking all sorts of morally questionable objectives. A couple of years ago, a small biotech company named Advanced Cell Technologies reported that it had successfully implanted human DNA into a cow's egg, and that that egg had successfully undergone a number of cell divisions into a viable blastocyst before it was destroyed.

It might come as a surprise to many that biotechnology is in a position to produce creatures that are part human and part animal, and that the law is indifferent as to whether it does so. If the Raelians or some other crackpot cult fund a human cloning experiment that results in the birth of a 15-pound, deformed infant, it is not clear that any U.S. law would have been broken—as long as they had not accepted any federal research dollars. Nor is it clear which government agency, if any, would have jurisdiction over such a case.

Congress is working to fix the latter problem with the introduction of a law, sponsored by Sen. Sam Brownback (R-Kan.), that bans human cloning altogether. This step is a positive one, insofar as it courageously seeks to reassert some control over the development of morally questionable technologies. It is important for society to draw certain red lines about the limits of our own self-modification. But in many prospective areas of research, total bans on particular technologies will not be the right answer.

As in the case of stem cells, most future developments in biotechnology will be ones in which positive developments will be intertwined with questionable ones. We will need to be able to make complex discriminations between the two. The same technology that will allow us to eliminate genes for cystic fibrosis or sickle-cell anemia from the human germ line may one day enable us to enhance competitively the intelligence or height of our children.

This implies the need for a standing regulatory structure that will be able to make and enforce distinctions between good and bad uses of biotechnology, on broader grounds than the safety criteria currently used by the Food and Drug Administration. And in the near term, it suggests the grounds for a possible compromise between those who want the current ban on federally funded research on stem cells to be lifted, and those who are concerned about encouraging further abortions. Why not accept the NIH guidelines for stem-cell research, but have them legislatively broadened to cover all research performed in the United States?

THE POLITICS OF CLONING

Eric Cohen

At various points in U.S. history, issues and events come along that make old ideologies obsolete, that make existing coalitions untenable, that make the contradictions within parties too pressing to ignore. When this happens, the old assumptions about politics no longer hold. The old battles begin to make little sense.

The genetic revolution—like slavery, civil rights and the Cold War before it—is such an event. In time, it may overshadow them all.

In the past few months, committees in both the House and the Senate have heard testimony on the now very real possibility of human cloning. Nearly all who testified agreed that initial attempts would bring horrific results—including spontaneous abortions, deformed children and grave dangers to the women who carried the first clones. With the exception of a few fringe cloning advocates, there is near universal consensus that cloning human beings with the intention of bringing them to birth ought not to be tried—at least not yet.

But the politics of cloning are complicated by the closely related issue of stem-cell research. Stem cells are undifferentiated cells that many scientists believe may someday cure many of our worst diseases and injuries. While there has been some promising research on stem cells harvested from adults, most scientists say the greatest promise is in embryonic cells—and that cloned embryos may be a key resource in this effort. Until now, this research has been barred from receiving federal funding.

This essay originally appeared in the *Los Angeles Times* on June 3, 2001.

But federal funding is only a small piece of a much larger set of moral and political issues: whether allowing human cloning to produce stem cells for research is the first step toward the cloning of newborns; whether the possible benefits of the new genetics justify the potential horrors; and whether we want to create whole new biotech industries that depend on a ready supply of human embryos, to be created, used and destroyed.

These questions are entangled in the politics of abortion—with the mostly pro-life Republicans treating embryos as sacred, and the mostly pro-choice Democrats seeing them as collections of cells that only have moral status if pregnant women choose to give them one. But the stakes, interests and complexity of the new genetics transcend, though certainly do not eclipse, the long-simmering abortion standoff. Stem cells and cloning, however significant, are only the beginning.

The mapping of the human genome raises the prospect of not just new genetic therapies for disease but genetic enhancements, or so-called germ-line interventions, that would affect all future generations. Eventually, the line between therapy and enhancement may become too difficult to draw—with genetic backwardness one day becoming the social equivalent of disease, and genetic equality becoming the next social egalitarian crusade.

Some research already underway, including the creation of genetic hybrids of human and animal embryos, raises the specter that our worst science-fiction nightmares may soon become possible. As of now, such experiments remain perfectly legal.

This brings us to the first wave of political and moral controversies in the new genetic age: Who regulates and what is regulated? What is allowed and what is banned? What guidance, if any, do prevailing ideologies and parties—conservatism and liberalism, Republicans and Democrats—supply for this next war of the gods?

There are many interests in this debate: religious conservatives and pro-life groups who see the new genetics as an affront to human dignity; biotech companies that stand to profit from the unregulated science of the future; reproductive-rights activists who see cloning as a personal choice that the government should not meddle with; and patient's rights groups who see in the new genetics a new source of hope.

The first faction, led by Sen. Sam Brownback (R-Kan.), wants to ban all human cloning (both "reproductive" and "research"). This group is made up of mostly Republicans and social conservatives, though it includes some members of the naturalist and environmentalist left, who remain skeptical of technological progress and open to government regulation. This pro-regulation movement has tried to separate the debate over human cloning from the politics of abortion, hoping that the nation's near-universal horror at the idea of live human clones will give them a substantial legislative victory.

Another new alliance of interests is that of reproductive-rights groups and biotech companies. This faction is willing to accept a temporary ban on reproductive cloning, on the grounds that the technology is still untested and unsafe. But it seeks to legalize and expand the use of cloned human embryos for medical research. It includes nearly all Democrats, who have been the most vigorous defenders of embryonic stem cells, and a sizable number of Republican capitalists, who believe free markets and technological progress are what defines their party, not religious conservatism and defense of the unborn.

The American middle, as usual, is somewhere in the uncertain in-between. Most Americans think abortion is wrong but should remain legal. They also believe that if embryos are available from abortions or in vitro-fertilization clinics, then using them for medical research is acceptable. They believe that the idea of live human clones is repugnant—just as most Americans once found the idea of test-tube babies repugnant. But they seem likely to tolerate the cloning of embryos for research if there are clear medical benefits of doing so. They want both moral seriousness and scientific progress, but seem to put initial moral concerns aside once experimental science shows itself to be safe.

Taken together, this coming battle will bring a sea change in American politics. If no ban or only a partial ban on human cloning is passed, then research will proceed in two directions. Those who seek to actually clone live humans will experiment on cloned human embryos to perfect the safety of the technique, so that they can eventually claim that human cloning is simply a reproductive choice. At the same time, scientists will attempt to use cloned embryos to develop revolutionary new medical therapies. If successful, they might well create a whole new industry and perhaps even a new economy: the economy of genes. This new capitalism will be, by its very nature, "pro-choice." The pro-life, pro-business, anti-regulation alliance that has long shaped American conservatism will become tenuous, if not impossible.

If the Republican Party casts its lot with the new capitalism, it will likely drive committed pro-lifers to revolt. The moral repugnance that spawned the influential conservative magazine *First Things* to declare "the end of democracy" in 1996 because of the institutionalization of abortion will return in earnest. Those who believe that the sanctity of life is violated—by the manufacture of human embryos, by the manipulation of man's genetic make-up, by the legalization of human cloning—may one day find America impossible.

But even if a ban on human cloning is achieved, it will remain precarious. Such controversial research is likely to go ahead in other countries, and if it is shown to be medically and economically successful, the floodgates of advocacy and protest from the biotech industry will be only that much more powerful. When and if this happens, the genetic capitalists will look to the Democrats, not the Republicans, as the anti-regulation, pro-industry party.

Where this is all heading is still unclear. That it will mean the end of liberalism and conservatism as we know them seems likely. There will be new coalitions, new divides. In the end, an America that often flees hard choices will have to make a very difficult one: between the new science, with its remarkable powers and unforeseen dangers, and the old faith or wisdom, with its humbling demands and veil-shrouded mysteries.

It seems likely that it will be our generation that must make this choice—a decision upon which the future of the American experiment, and perhaps the fate of mankind itself, may ultimately hinge. That we are unprepared is the understatement of the pre-genetic age.

LETTER TO HHS SECRETARY TOMMY G. THOMPSON

Senator Orrin G. Hatch

Dear Mr. Secretary:

I am writing to express my views regarding federal funding of biomedical research involving human pluripotent embryonic stem cells. After carefully considering the issues presented, I am persuaded that such research is legally permissible, scientifically promising, and ethically proper.

***

At the outset, let me be clear about one of my key perspectives as a legislator: I am pro-family and pro-life. I abhor abortion and strongly oppose this practice except in the limited cases of rape, incest, and to protect the life of the mother. While I respect those who hold a pro-choice view, I have always opposed any governmental sanctioning of a general abortion on demand policy. In my view, the adoption of the Hyde Amendment wisely restricts taxpayer financed abortions. Moreover, because of my deep reservations about the Supreme Court's decision in *Roe* v. *Wade*, I proposed—albeit unsuccessfully—an amendment to the Constitution in 1981 that would have granted to the states and Congress the power to restrict or even outright prohibit abortion.

In 1992, I led the Senate opposition to fetal tissue research that relied upon cells from induced abortions. I feared that such research would be used to justify abortion or lead to additional abortions. It was my under-

Senator Orrin G. Hatch (R) is a member of the U.S. Senate from Utah. He was elected to the Senate in 1976 and reelected in 1982, 1988, 1994, and 2000. He is one of the most prominent and outspoken "pro-life" advocates for embryonic stem cell research. Included here are selections from his June 13, 2001, letter to Secretary of Health and Human Services Tommy G. Thompson.

standing that tissue from spontaneous abortions and ectopic pregnancies could provide a sufficient and suitable supply of cells. Unfortunately, experts did not find these sources of cells as adequate for their research needs. . . .

I am proud of my strong pro-life, anti-abortion record. I commend the Bush Administration for its strong pro-life, pro-family philosophy. In my view, research on stem cells derived from embryos first created for, but ultimately not used in, the process of in vitro fertilization, raises questions and considerations fundamentally different from issues attendant to abortion. As I evaluate all these factors, I conclude that this research is consistent with bedrock pro-life, pro-family values. I note that our pro-life, pro-family Republican colleagues, Senators Strom Thurmond and Gordon Smith, as well as former Senator Connie Mack, support federal funding of embryonic stem cell research. It is my hope that once you have analyzed the issues, you will agree with us that this research should proceed.

***

While society must take into account the potential benefits of a given technological advance, neither scientific promise nor legal permissibility can ever be wholly sufficient to justify proceeding down a new path. In our pluralistic society, before the government commits taxpayer dollars or otherwise sanctions the pursuit of a novel field of research, it is imperative that we carefully examine the ethical dimensions before moving, or not moving, forward.

I would hope there is general agreement that modern techniques of in vitro fertilization are ethical and benefit society in profound ways. I have been blessed to be the father of six children and the grandfather of nineteen grandchildren. Let me just say that whatever success I have had as a legislator pales in comparison to the joy I have experienced from my family in my roles of husband, father, and grandfather. Through my church work, I have counseled several young couples who were having difficulty in conceiving children. I know that IVF clinics literally perform miracles every day. It is my understanding that in the United States over 100,000 children to date have been born through the efforts of IVF clinics.

Intrinsic with the current practice of IVF-aided pregnancies is the production of more embryos than will actually be implanted in hopeful mothers-to-be. The question arises as to whether these totipotent embryonic cells, now routinely and legally discarded—amid, I might add, no great public clamor—should be permitted to be derived into pluripotent cells with non-federal funds and then be made available for research by federal or federally-supported scientists?

Cancer survivor and former Senator, Connie Mack, recently explained his perspective on the morality of stem cell research in a *Washington Post* op-ed piece:

> It is the stem cells from surplus IVF embryos, donated with the informed consent of couples, that could give researchers the chance to move embryonic stem cell research forward. I believe it would be wrong not to use them to potentially save the lives of people. I know that several members of Congress who consider themselves to be pro-life have also come to this conclusion.

Senator Mack's views reflect those of many across our country and this perspective must be weighed before you decide.

Among those opposing this position is Senator Brownback, who has forcefully expressed his opinion:

> The central question in this debate is simple: Is the embryo a person, or a piece of property? If you believe . . . that life begins at conception and that the human embryo is a person fully deserving of dignity and the protection of our laws, then you believe that we must protect this innocent life from harm and destruction.

While I generally agree with my friend from Kansas on pro-life, pro-family issues, I disagree with him in this instance. First off, I must comment on the irony that stem cell research—which under Senator Brownback's construction threatens to become a charged issue in the abortion debate—is so closely linked to an activity, in vitro fertilization, that is inherently and unambiguously pro-life and pro-family.

I recognize and respect that some hold the view that human life begins when an egg is fertilized to produce an embryo, even if this occurs *in vitro* and the resulting embryo is frozen and never implanted *in utero*. To those with this perspective, embryonic stem cell research is, or amounts to, a form of abortion. Yet this view contrasts with statutes, such as Utah's, which require the implantation of a fertilized egg before an abortion can occur.

[The question is] whether a frozen embryo stored in a refrigerator in a clinic is really equivalent to an embryo or fetus developing in a mother's womb? To me, a frozen embryo is more akin to a frozen unfertilized egg or frozen sperm than to a fetus naturally developing in the body of a mother. In the case of in vitro fertilization, extraordinary human action is required to initiate a successful pregnancy while in the case of an elective abortion an intentional human act is required to terminate pregnancy. These are polar opposites. The purpose of in vitro fertilization is to facilitate life while abortion denies life. Moreover, as Dr. Louis Guenin has argued: "If we spurn [embryonic stem cell research] not one more baby is likely to be born." I find the practice of attempting to bring a child into the world through in vitro fertilization to be both ethical and laudable and distinguish between elective abortion and the discarding of frozen embryos no longer needed in the in vitro fertilization process.

In evaluating this issue, it is significant to point out that no member of the United States Supreme Court has ever taken the position that fetuses, let alone embryos, are constitutionally protected persons. To do so would be to thrust the courts and other governmental institutions into the midst of some of the most private of personal decisions. For example, the use of contraceptive devices that impede fertilized eggs from attaching onto the uterine wall could be considered a criminal act. Similarly, the routine act of discarding "spare" frozen embryos could be transformed into an act of murder.

As much as I oppose partial birth abortion, I simply cannot equate this offensive abortion practice with the act of disposing of a frozen embryo in the case where the embryo will never complete the journey toward birth. Nor, for example, can I imagine Congress or the courts somehow attempting to order every "spare" embryo through a full term pregnancy.

Mr. Secretary, I greatly appreciate your consideration of my views on this important subject. I only hope that when all the relevant factors are weighed both you and President Bush will decide that the best course of action for America's families is to lead the way to a possible new era in medicine and health by ordering that this vital and appropriately regulated research proceed.

FOR A TOTAL BAN ON HUMAN CLONING

J. Bottum and William Kristol

About the horror of creating human beings by cloning, there is wide agreement among the American people—and in Congress as well.

But about the extent of the necessary ban on cloning—whether it must outlaw all human cloning or only cloning that aims explicitly at bringing a cloned child to birth—controversy has arisen. It was on this issue that Congress's attempt to prohibit cloning foundered in 1998. And now, in 2001, Congress faces the issue once again with a pair of competing bills. The first is a complete ban on the cloning of human beings, sponsored in the House by Florida Republican Dave Weldon and Michigan Democrat Bart Stupak, and in the Senate by Kansas Republican Sam Brownback. The second is a partial ban, prohibiting for ten years only so-called "reproductive" cloning, sponsored by Pennsylvania Republican James Greenwood and backed by the biotech lobby.

The Weekly Standard has editorialized before about the practical impossibility and moral fecklessness of trying to ban cloning only for certain purposes. Greenwood's bill permits and even encourages scientists to create cloned embryos, and then attempts to prevent them from inserting those embryos into a womb. This is unenforceable, since only a judicially ordered abortion could eliminate the result of violating such a law. It is unprosecutable, since it would require an unattainable knowledge of a scientist's intention in creating a clone. And it is unethical, since it would establish, for the first time in federal statutes, a class of embryos that it is a crime *not* to de-

This essay originally appeared in the *Weekly Standard* on July 2/9, 2001.

stroy, a felony *not* to treat as anything except disposable tissue. A ban solely on "reproductive" cloning will prove nothing more—and nothing less—than a license to clone. There is, in truth, only one anti-cloning proposal before Congress: the Weldon-Stupak bill.

On June 19 and June 20, a pair of hearings were held in the House of Representatives—one by a Judiciary subcommittee, chaired by Lamar Smith, and the other by an Energy and Commerce subcommittee, chaired by Michael Bilirakis—on the Weldon-Stupak and Greenwood bills. From a political perspective, the most important testimony at these hearings came from Claude Allen, the deputy secretary of the Department of Health and Human Services, who presented, for the first time in an official setting, the Bush administration's position: strong support for a ban on "any and all attempts to clone a human being" and complete rejection of pseudo-compromises like the Greenwood bill.

"The administration favors the passage of specific legislation to prohibit the cloning of a human being," Allen told the Energy and Commerce subcommittee. The Weldon-Stupak bill, he added, "is consistent with [Health and Human Services] secretary [Tommy] Thompson's and the president's views."

The administration deserves credit for taking an unequivocal position against all human cloning, in the face of pressure from some in the biotech industry. This issue, however, will require personal leadership from President Bush—and sooner rather than later. He will have to help educate the American public about cloning and work to move the ban through the House and Senate. (The House leadership intends to seek a vote on Weldon-Stupak as early as next month.)

The president will be aided in his task by the extraordinary testimony of many distinguished witnesses at last week's hearings. They ranged across the political spectrum, and the moral seriousness and eloquence they brought to the halls of Congress was striking.

Consider the testimony of social philosopher Jean Bethke Elshtain of the University of Chicago:

> The path down which we are headed unless we intervene now to stop human cloning is one that will deliver harm in abundance—and harm that can be stated clearly and decisively *now*—whereas any potential benefits are highly speculative and likely to be achievable through less drastic and damaging methods, in any case. The harms, in other words, are known—not a matter of speculation—whereas the hypothesized benefits are a matter of conjecture, in some cases rather far-fetched conjecture. . . .

> The hope of genetic fundamentalists is that we can increasingly control for that which is deemed desirable and eliminate that which is not. The aim in all this is not to prevent devastating illnesses but precisely to reflect and to reinforce certain societal prejudices in and through genetic selection. There is a word for this so-called "genetic enhancement." That word is eugenics. Human cloning belongs to this eugenics project.

The always-thoughtful Francis Fukuyama of Johns Hopkins added:

> Cloning represents the opening wedge for a series of future technologies that will permit us to alter the human germline and ultimately to design people genetically. . . . It is therefore extremely important that Congress act legislatively at this point to establish the principle that our democratic political community is sovereign and has the power to control the pace and scope of such technological developments. . . .
>
> Opponents of a legislative ban frequently argue that such a ban would be rendered ineffective by the fact that we live in a globalized world in which any attempt to regulate technology by sovereign nation-states can easily be sidestepped by moving to another jurisdiction. . . . This is part of a larger widespread belief that technological advance should not and cannot be stopped.
>
> I believe that this is a fundamentally flawed argument. In the first place, it is simply not the case that the pace and scope of technological advance cannot be controlled politically. . . . Second, to argue that no national ban or regulation can precede an international agreement on the subject is to put the cart before the horse. Regulation never starts at an international level: Nation-states have to set up enforceable rules for their own societies before they can even begin to think about international rules. . . .
>
> If we can establish a general consensus among civilized nations that human cloning is unacceptable, we will then have a range of traditional diplomatic and economic instruments at our disposal to persuade or pressure countries outside that consensus to join.

Another witness in support of Weldon-Stupak was Judy Norsigian, executive director of the Boston Women's Health Book Collective and coauthor of the latest edition of *Our Bodies, Ourselves,* the most widely distributed feminist text in America. As she noted, her organization has "a long track record" of support for legalized abortion. But, as she explained,

> Cloning advocates are seeking to appropriate the language of reproductive rights to support their case. This is a travesty. There is an immense difference between seeking to end an unwanted pregnancy and seeking to create a genetic duplicate human being. . . .

> The [Boston Women's Health Book Collective] joins many other national and international organizations in calling for a universal ban on human reproductive cloning. To allow the creation of human clones would open the door to treating our children like manufactured objects. It would violate deeply and widely held values concerning human individuality and dignity. It would pave the way for unprecedented new forms of eugenics. And it would serve no justifiable purpose.

Important testimony was also offered by Alexander Capron, a Clinton appointee to the National Bioethics Advisory Commission and professor of law at the University of Southern California, by Gerard Bradley of the University of Notre Dame law school, by Richard Doerflinger of the National Conference of Catholic Bishops, and by Stuart Newman, professor of cell biology and anatomy at New York Medical College and co-founder of the Council for Responsible Genetics. But perhaps the most ringing words came from Leon Kass of the University of Chicago:

> As has been obvious for some time, new biotechnologies are providing powers to intervene in human bodies and minds in ways that go beyond the traditional goals of healing the sick, to threaten fundamental changes in human nature and the meaning of our humanity. These technologies have now brought us to a crucial fork in the road, where we are compelled to decide whether we wish to travel down the path that leads to the Brave New World. That, and nothing less, is what is at stake in your current deliberations about whether we should tolerate the practice of human cloning. . . .
>
> [Cloning] constitutes unethical experimentation on the child-to-be, subjecting him or her to enormous risks of bodily and developmental abnormalities. It threatens individuality. . . . It confuses identity. . . . It represents a giant step toward turning procreation into manufacture. . . . And it is a radical form of parental despotism and child abuse. . . . A majority of members of Congress, I believe, are, like most Americans, opposed to human cloning. But opposition is not enough. For if Congress does nothing about it, we shall have human cloning, and we shall have it soon. Congress' failure to try to stop human cloning—and by the most effective means—will in fact constitute its tacit approval.

There was a revealing moment in the question period that followed the testimony—one of those brief flashes that expose our current situation—when Congressman Ted Strickland of Ohio complained, "We should not allow theology, philosophy, or politics to interfere with the decision we make on this issue."

If we do not allow theology to inform our political decisions, if we do not allow philosophy, if we do not allow even politics, what remains? Only the final paralysis of government and the final paralysis of thought—which is to say, the inevitability of cloned human beings and the degradation of human liberty and dignity.

CELL OUT

William Saletan

A new camp has emerged in the debate over abortion and fetal tissue. A group of Republicans led by Sen. Orrin Hatch of Utah, Sen. Gordon Smith of Oregon, and former Sen. Connie Mack of Florida is lobbying President Bush to support federally funded research on human embryonic stem cells. Right-to-life groups oppose such research on the grounds that the harvesting of these cells kills the embryos, which are left over from in vitro fertilizations. The Hatch-Mack-Smith camp, however, says it's possible to be both "pro-life" and "pro-stem cells." The senators—let's call them pro-pros—are trying to give Bush a rationale for stem-cell research that doesn't entail acceptance of abortion. But if Bush embraces their principles, he'll be what they are: functionally pro-choice. Let's look at those principles.

1. Personhood is situational. The pro-pros concede that if personhood begins at conception, it's immoral to destroy IVF embryos by taking cells from them. But personhood doesn't begin at conception, they say. It begins when the embryo is implanted in a womb. "To me, a frozen embryo is more akin to a frozen unfertilized egg or frozen sperm than to a fetus naturally developing in the body of a mother," Hatch wrote in a letter he forwarded to Bush last month. In an ABC interview, Hatch flatly declared, "Human life begins in the mother's womb, not in a petri dish or a refrigerator." At a recent hearing, Smith added, "If you have a stem cell in a petri dish and you keep it

William Saletan is a senior writer at *Slate.com* and is writing a book on abortion and American politics. This essay originally appeared in *Slate.com* on July 12, 2001.

there for 50 years, you'll end up with a stem cell in a petri dish. And until you place that in a woman, you are not going to create a life."

On this theory, the value of human life, like the value of a house, is determined by location, location, location. Whether it's an egg or an embryo is less important than whether it's in a womb or a freezer. But location, unlike fertilization, is easily reversed. Take the embryo out of the woman, and it ceases to be a person. The pro-pro response to this objection is that you can't take the embryo out, since it's already a person. But if location, not fertilization, is what makes the embryo a person, then doesn't changing its location change its status? You don't have to kill it; you can just put it in a dish or a freezer. That's what happens to eggs in IVF, and the pro-pro senators have no problem with it. Is it OK to do this to an egg but not to an embryo? Why? The only difference is fertilization, which the pro-pros have dismissed as the standard of personhood. Having stipulated that a frozen embryo is more like a frozen egg than like an embryo in the womb, they have no grounds to complain. The adjective trumps the noun.

In his letter, Hatch noted that he "helped lead the effort to outlaw partial birth abortion." He concluded, "I simply cannot equate this offensive abortion practice with the act of disposing of a frozen embryo in the case where the embryo will never complete the journey toward birth." But wait a minute. The first step in partial-birth abortion is induced delivery. If the frozen embryo is a non-person because it "will never complete the journey toward birth," then why isn't the same true of a prematurely delivered five-month fetus? You can't shove the fetus back into the woman. Like the frozen egg or embryo, it's no longer "naturally developing in the body of a mother." In fact, by Hatch's standard, a five-month fetus has less claim to personhood than a frozen embryo does, because whereas the completion of the embryo's "journey toward birth" is unlikely (depending on whether some woman will consent to have it implanted), the completion of the fetus's journey, given current technology, is impossible.

Hatch says his position on stem cells should be accepted as pro-life because his position on partial-birth abortion is pro-life. But the two positions aren't just irreconcilable. They're exact opposites. On the one hand, Hatch proposes to ban the one abortion procedure in which the fetus is removed from the womb intact and then dismembered. Like other advocates of this selective ban, he justifies it precisely on the grounds that the abortion is performed in this sequence. On the other hand, Hatch tells Bush that it's OK to dismember an IVF embryo, since the egg from which that embryo was created has already been removed from and fertilized outside the womb.

The criteria are exactly the same—location and sequence—but the position is reversed.

2. It's OK to dismember an embryo if it's unwanted. According to Hatch, surplus IVF embryos "are going to be thrown away. They are going to be discarded. They're going to be killed, if you will. Why can't we take the pluripotent cells from them and utilize them for the best benefits of mankind?" One of the telling oddities of the stem-cell debate is this constant use of the passive voice. Somebody else will do the killing; all we can do is make the best of it. This is precisely the attitude of resigned relativism that pro-lifers despise in pro-choicers. When pro-choicers say it's acceptable to get an abortion if the baby is unwanted and would die or be abandoned, pro-lifers reply that the baby can and must be given a good home. The pro-life outlook is idealistic: Tragedy isn't inevitable, and instead of watching it, you can do something to avert it. From that point of view, the government should preserve IVF embryos and facilitate their implantation in women who want to adopt them, as 14 House members proposed in a June 28 letter to Bush.

3. Keep embryo dismemberment safe and legal. "Policymakers should also consider another advantage of public funding of stem cell research as opposed to leaving this work beyond the reach of important federal controls," Hatch wrote. "Federal funding will encourage adherence to all of the safeguards outlined above by entities conducting such research even when a particular research project is conducted solely with private dollars." Or, as pro-choicers would put it: If you try to stop abortions, you'll just drive them underground.

4. Embryo dismemberment is pro-life and pro-family because it prolongs lives and helps families. "The most pro-life position would be to help people who suffer from these maladies" that could be cured through stem-cell research, Hatch told the *Washington Post*. "This is about extending life, facilitating life," he told ABC. Smith took the point further: "Part of being pro-life is helping the living." In his letter to Bush, Hatch added, "It is also worth noting in the pro-family context that stem cell research is of particular interest to pediatricians. . . . [T]he knowledge gained through biomedical research can be harnessed for critical pro-life, pro-family purposes. When one of our loved ones is stricken by illness, the whole family shares in the suffering."

These flexible interpretations of "pro-life" and "pro-family" aren't original. Pro-choicers commonly argue that abortion is "pro-life" when it helps the living and "pro-family" when it helps a family. In the partial-birth debate, they reason that late-term abortions are pro-life and pro-family because such abortions ensure that women with abnormal pregnancies won't require hysterectomies, and therefore those women will be able to bring other lives into the world and raise families. The catch, of course, is that the "life" that's helped isn't the one that's taken. By this utilitarian logic, the Chinese government's policy of executing prisoners and harvesting their organs for transplant is pro-life. Such lethal utilitarianism is exactly what the right-to-life movement was founded to resist. It's easy to be "pro-life." What's hard is defending the *right* to life.

5. Embryo dismemberment is the parents' choice. In a *Post* op-ed two months ago, Mack pointed out that the stem cells sought for research had been "donated with the informed consent of couples." Quoting those words in his letter, Hatch advised Bush, "Senator Mack's views reflect those of many across our country and this perspective must be weighed before you decide." Hatch concluded: "It is significant to point out that no member of the United States Supreme Court has ever taken the position that fetuses, let alone embryos, are constitutionally protected persons. To do so would be to thrust the courts and other governmental institutions into the midst of some of the most private of personal decisions."

It's hard to imagine how anyone who wrote those words could truly believe in an unborn child's right to life. That doesn't mean the pro-pros are insincere about abortion or stem cells. It just means they haven't yet put two and two together. This is how morality changes. An issue troubles your conscience, and you look for ways to revise your thinking on that issue without upsetting your thinking on other issues. But gradually, you come to see that it's all connected. That's the nature of thinking. Eventually, you realize that you've lost faith in what you used to believe. The pro-pros set out to change the president's mind. They'll end up changing their own.

OF MISSILE DEFENSE AND STEM CELLS

Eric Cohen

Among the issues in American politics that inspire the most ideological fervor these days, stem cells and missile defense are at the top of the list. Missile defense has a long history: To conservative Republicans, it is a fixture of the Reagan legacy, of American strength, independence, and nuclear realism in the post–Cold War world. To liberal Democrats, missile defense is destabilizing, hegemonic, unworkable, and unwise. It will provoke a new arms race and a new age of nuclear brinkmanship. Besides which, terrorists can always attack us with nuclear car bombs anyway.

The issue of stem cells is new—a continuation of the moral and political divide over abortion, but with perhaps even greater complexity and significance. Pro-lifers see research on embryonic stem cells as involving the utilitarian destruction of the unborn. And they see it as the gateway to the darker, more ambitious modern genetic project of designing our descendants and challenging our mortality. Among the supporters of this research, by contrast, the pro-capitalists and many "soft" pro-lifers foresee staggering benefits that far outweigh any associated evil. The pro-choicers see no evil at all, only a great humanitarian opportunity to extend individual health and autonomy.

What is interesting, though, are the parallel claims and counterclaims made by those who advocate and those who reject these two emerging technologies. The advocates proclaim: If we lift the respective bans—the ABM treaty and the NIH regulations barring federal funding of embryonic stem cell research—technological miracles will follow. The skeptics proclaim: These technologies

This essay originally appeared in the *Weekly Standard* on July 16, 2001.

are untested, immoral, and irresponsible. On each issue, the pro-technology faction asserts not only the virtue of deploying either missile defense or stem cells, but the *necessity* of doing so—lest terrorists attack us or diseases kill us.

And usually—here is perhaps the most interesting point of all—the advocates of one technology reject the other. That is, missile-defense hawks, who tend to be conservatives, are usually stem cell doves; stem cell hawks, who tend to be liberals, are usually missile-defense doves. There are exceptions, of course, but the discontinuity is common enough to be worth considering.

Perhaps not surprisingly, the two subjects are seldom discussed in the same political breath. But the relationship between the politics of nuclear weapons and the politics of the new biology is fundamental: Both stem cell research and missile defense force concrete judgments about whether modern technology enhances life or threatens it, whether it expands freedom or destroys it. Both inspire grand fears about where modern technology is leading us. Both raise questions about whether we can control what we create and what we are, and about whether such control is desirable, undesirable, or tragically necessary.

For conservatives in particular, these issues present a riddle—especially for those who seek both to augment American greatness and power, on the one hand, and to demand of the nation a technological reticence, a reverence for the unmanageable mystery of creation, and a spirit of restraint and acceptance in the face of suffering, on the other. These conservatives seem to want a "just hegemony" in international affairs, built on America's will to set the world right. But when it comes to the irrationalities and inevitability of suffering, disability, sickness, and death, they ask the nation to adopt, as bioethicist Gilbert Meilaender eloquently puts it, "the posture of one who waits, who knows his fundamental neediness and dependence."

In short, they seek both the posture of the heroic statesman and the posture of man as witness. American conservatism, at its best, cultivates both, in deference to a paradox inherent in the human condition. But politically, it is not enough simply to lift the ban on weapons-builders and maintain the ban on medical researchers, declaring oneself pro-defense and pro-life. Rather, this conservative disposition must be seen to make sense.

For the fact is, as Meilaender and others have suggested, the philosophical problems posed by our willingness to fight just wars and our desire to cure diseases are not very different. Both endeavors confront us with seemingly impossible questions: When may we take life to affirm life? Can embryos ever be justly sacrificed to help the sick and dying? Are discarded embryos acceptable "collateral damage" in the war against disease? When does courage require of us that we endure our fate, and when that we exert the will to set the world right? How and when should we use power to extend the "pursuit of happiness," be it American power overseas or medical power at home? In short: How much goodness and how much justice can men achieve here and now? And when does wisdom require the heroic acceptance of tragedy, forbearance rather than "progress" and "solutions"?

In my view, building a missile defense system and halting all embryonic stem cell research are the moral and realistic choices. But those who adopt this set of positions must recognize the grand wagers they rest on: namely, that a nuclear attack is possible but not inevitable; that missile defense is workable and will deter our enemies rather than embolden them; that the biological quest to overcome suffering—to set the world right by ending disease and perfecting imperfection—is somehow misguided; and that the further down this path we go, the less able we will be to accept, endure, and redeem our mortality, or to preserve the dignity of the "imperfect" among us, which in the end means all of us. This treating of life as a problem to be solved has given us the modern capacity to cure disease, but also our increasing penchant for euthanasia, assisted suicide, mass Prozac, and selective abortions.

Certainly, these two conservative positions (pro-missile defense and American power, anti-embryonic stem cell research) are difficult to reconcile—the one a mobilization of modern technology, the other a call to rein it in. To acknowledge the force of the opposing views—the futility of fighting nuclear weapons with more weapons, the rightness of extending the lives of the sick and the dying even at the cost of destroying "mere cells"—is a necessity.

Perhaps the answer, if there is one, lies in America's exceptional conservatism, which in the past has inspired both the will to fight tyranny and the wisdom to acknowledge man's limits, and hence his longing for transcendent redemption or justice. To ask comfortable citizens to give their lives defending freedom around the world; to ask the sick and dying to love the mystery of life more than their own lives—both require a courageous commitment to something larger than self-interest. For a purely political conservatism oriented toward giving the voters what they want, such demands are a losing strategy. For a philosophically grounded conservatism willing to risk demanding from people the sacrifices of which they are capable, these issues are an exceptional opportunity.

TESTIMONY

U.S. Senate

Michael D. West

I am pleased to testify today in regard to the new opportunities and challenges associated with human embryonic stem (ES) cell and nuclear transfer (NT) technologies. I will begin by describing the bright promise of these twin and interrelated technologies and then attempt to correct some misunderstandings relating to their application in medicine.

It may be useful to point out that I think of myself as pro-life in that I have an enormous respect for the value of the individual human life. Indeed, in my years following college I joined others in the protest of abortion clinics. My goal was not to send a message to women that they did not have the right to choose. My intent was simply to urge them to reconsider the destruction of a developing human being. Despite my strong convictions about the value of the individual human life, in 1995 I organized the collaboration between Geron Corporation and the laboratories of Dr. James Thomson and John Gearhart to isolate human embryonic stem cells and human embryonic germ cells from human embryos and fetuses respectively. My reasons were simple. These technologies are entirely designed to be used in medicine to alleviate human suffering and to save human life. They are, in fact, pro-life. The opponents that argue they destroy human lives are simply and tragically mistaken. Let me explain why this is the case.

Michael D. West is a biologist and president and CEO of Advanced Cell Technology, Inc., a biotechnology company in Worcester, Massachusetts. He is also the founder of Geron Corporation, one of the major patent holders of embryonic stem cells. Selections from testimony before the U.S. Senate Committee on Appropriations, Subcommittee on Labor, Health, and Human Services, July 18, 2001.

HUMAN ES CELLS

We are composed of trillions of individual living cells, glued together like the bricks of a building to construct the organs and tissues of our body. The cells in our bodies are called "somatic cells" to distinguish them from the "germ line," that is, the reproductive cells that connect the generations. We now know that life evolved from such single-celled organisms that dominated all life some one billion years ago.

Therefore, in answer to the question of when life begins, we must make a crucial distinction. Biological life, that is to say, "cellular life" has no recent beginnings. Our cells are, in fact, the descendants of cells that trace their beginnings to the origin of life on earth. However, when we speak of an individual human life, we are speaking of the communal life of a multicellular organism springing from the reproductive lineage of cells. The individual human life is a body composed of cells committed to somatic cell lineages. All somatic cells are related in that they originate from an original cell formed from the union of a sperm and egg cell.

The fertilization of the egg cell by a sperm leads to a single cell called the "zygote." From this first cell, multiple rounds of cell division over the first week result in a microscopic ball of cells with very unusual properties. This early embryo, called the "preimplantation embryo," has not implanted in the uterus to begin a pregnancy. It is estimated that approximately 40% of preimplantation embryos formed following normal human sexual reproduction fail to attach to the uterus and are naturally destroyed as a result.

From the above it should be clear that at the blastocyst stage of the preimplantation embryo, no body cells of any type have formed, and even more significantly, there is strong evidence that not even the earliest of events in the chain of events in somatic differentiation have been initiated. A simple way of demonstrating this is by observing subsequent events.

Should the embryo implant in the uterus, the embryo, at approximately 14 days post fertilization will form what is called the primitive streak, this is the first definition that these "seed" cells will form an individual human being as opposed to the forming of two primitive streaks leading to identical twins. Rarely two primitive streaks form that are not completely separated leading to conjoined or Siamese twins. In addition, rarely, two separately fertilized egg cells fuse together to form a single embryo with two different cell types. This natural event leads to a tetragammetic chimera, that is a single human individual with some of the cells in their body being male from the original male embryo, and some cells being female from the original female embryo. These and other simple lessons in embryology teach us that despite the dogmatic assertions of some theologians, the evidence is decisive in support of the position that an individual human life, as opposed to merely cellular life, begins with the primitive streak, (i.e., after 14 days of development). Those who argue that the preimplantation embryo is a person are left with the logical absurdity of ascribing to the blastocyst personhood when we know, scientifically speaking, that no individual exists (i.e., the blastocyst may still form identical twins).

Human ES cells are nothing other than ICM cells grown in the laboratory dish. Because these are pure stem cells uncommitted to any body cell lineage, they may greatly improve the availability of diverse cell types urgently needed in medicine. Human ES cells are unique in that they stand near the base of the developmental tree. These cells are frequently designated "totipotent" stem cells, meaning that they are potentially capable of forming any cell or tissue type needed in medicine. These differ from adult stem cells that are "pluripotent" that is, capable of forming several, but only a limited number, of cell types. An example of pluripotent adult stem cells are the bone marrow stem cells now widely used in the treatment of cancer and other life-threatening diseases.

Some have voiced objection to the use of human ES cells in medicine owing to the source of the cells. Whereas the use of these new technologies has already been carefully debated and approved in the United Kingdom, the United States lags disgracefully behind. I would like to think it is our goodness and our kindness as a people that generates our country's anxieties over these new technologies. Indeed, early in my life I might have argued that since we don't know when a human life begins, it is best not to tamper with the early embryo. That is to say, it is better to be safe than sorry. I believe many U.S. citizens share this initial reaction. But, with time the facts of human embryology and cell biology will be more widely understood. As the Apostle Paul said: "When I was a child, I spake as a child, I understood as a child, I thought as a child: but when I became a man, I put away childish things." (I Cor 13:11) In the same way it is absolutely a matter of life and death that policy makers in the United States carefully study the facts of human embryology and stem cells. A child's understanding of human reproduction simply will not suffice and such ignorance could lead to disastrous consequences.

With appropriate funding of research, we may soon learn to direct these cells to become vehicles of lifesaving potential. We may, for instance, become able to produce neurons for the treatment of Parkinson's disease and spinal cord injury, heart muscle cells for heart failure, cartilage for arthritis and many others as well. This research has great potential to help solve the first problem of tissue availability, but the technologies to direct these cells to become various cell types in adequate quantities remains to be elucidated. Because literally hundreds of cell types are needed, thousands of academic research projects need to be funded, far exceeding the resources of the biotechnology industry.

As promising as ES cell technology may seem, it does not solve the remaining problem of histocompatibility. Human ES cells obtained from embryos derived during in vitro fertilization procedures, or from fetal sources, are essentially cells from another individual (allogeneic). Several approaches can be envisioned to solve the problem of histocompatibility. One approach would be to make vast numbers of human ES cell lines that could be stored in a frozen state. This "library" of cells would then offer varied surface antigens, such that the patient's physician could search through the library for cells that are as close as possible to the patient. But these would likely still require simultaneous immunosuppression that is not always effective. In addition, immunosuppresive therapy carries with it increased cost, and the risk of complications including malignancy and even death.

Another theoretical solution would be to genetically modify the cultured ES cells to make them "universal donor" cells. That is, the cells would have genes added or genes removed that would "mask" the foreign nature of the cells, allowing the patient's immune system to see the cells as self. While such technologies may be developed in the future, it is also possible that these technologies may carry with them unacceptably high risks of rejection or other complications that would limit their practical utility in clinical practice.

Given the seriousness of the current shortage of transplantable cells and tissues, the FDA has demonstrated a willingness to consider a broad array of options including the sourcing of cells and indeed whole organs from animals (xenografts) although these sources also pose unique problems of histocompatibility. These animal cells do have the advantage that they have the potential to be genetically engineered to approach the status of "universal donor" cells, through genetic engineering. However as described above, no simple procedure to confer such universal donor status is known. Most such procedures are still experimental and would likely continue to require the use of drugs to hold off rejection, drugs that add to health care costs, and carry the risk of life-threatening complications.

THERAPEUTIC CLONING

An extremely promising solution to this remaining problem of histocompatibility would be to create human ES cells genetically identical to the patient. While no ES cells are known to exist in a developed human being and are therefore not available as a source for therapy, such cells could possibly be obtained through the procedure of somatic cell nuclear transfer (NT), otherwise known as cloning technology. In this procedure, body cells from a patient would be fused with an egg cell that has had its nuclear DNA removed. This would theoretically allow the production of a blastocyst-staged embryo genetically identical to the patient that could, in turn, lead to the production of ES cells identical to the patient. In addition, published data suggests that the procedure of NT can "rejuvenate" an aged cell restoring the proliferative capacity inherent in cells at the beginning of life. This could lead to cellular therapies with an unprecedented opportunity to improve the quality of life for an aging population.

The use of somatic cell nuclear transfer for the purposes of dedifferentiating a patient's cells and obtain[ing] autologous undifferentiated stem cells has been designated "Therapeutic Cloning" or alternatively, "Cell Replacement by Nuclear Transfer." This terminology is used to differentiate this clinical indication from the use of NT for the cloning of a child that in turn is designated "Reproductive Cloning." In the United Kingdom, the use of NT for therapeutic cloning has been carefully studied by their Embryology Authority and formally approved by the Parliament.

ETHICAL CONSIDERATIONS

Ethical debates often center over two separate lines of reasoning. Deontological debates are, by nature, focused on our duty to God or our fellow human being. Teleological arguments focus on the question of whether the ends justify the means. Most scholars agree that human ES cell technology and therapeutic cloning offer great pragmatic merit, that is, the teleological arguments in favor of ES and NT technologies are quite strong. The lack of agreement, instead, centers on the deontological arguments relating to the rights of the blastocyst embryo and our duty to protect the individual human life.

I would argue that the lack of consensus is driven by a lack of widespread knowledge of the facts regarding the origins of human life on a cellular level and human life on a somatic and individual level. So the question of when does life begin, is better phrased "when does an individual human life begin." Some dogmatic individuals claim with the same certainty the Church opposed Galileo's claim that the earth is not the center of the universe, that an individual human life begins with the fertilization of the egg cell by the sperm cell. This is superstition, not science. The belief that an individual human being begins with the fertilization of the egg cell by the sperm cell is without basis in scientific fact or, for that matter, without basis in religious tradition.

All strategies to source human cells for the purposes of transplantation have their own unique ethical problems. Because developing embryonic and fetal cells and tissues are "young" and are still in the process of forming mature tissues, there has been considerable interest in obtaining these tissues for use in human medicine. However, the use of aborted embryo or fetal tissue raises numerous issues ranging from concerns over increasing the frequency of elected abortion to simple issues of maintaining quality controls standards in this hypothetical industry. Similarly, obtaining cells and tissues from living donors or cadavers is also not without ethical issues. For instance, an important question is, "Is it morally acceptable to keep 'deceased' individuals on life support for long periods of time in order to harvest organs as they are needed?"

The implementation of ES-based technologies could address some of the ethical problems described above. First, it is important to note that the production of large numbers of human ES cells would not in itself cause these same concerns in accessing human embryonic or fetal tissue, since the resulting cells have the potential to be grown for very long periods of time. Using only a limited number of human embryos not used during in vitro fertilization procedures, could supply many millions of patients if the problem of histocompatibility could be resolved. Second, in the case of NT procedures, the patient may be at lower risk of complications in transplant rejection. Third, the only human cells used would be from the patient. Theoretically, the need to access tissue from other human beings could be reduced.

Having knowledge of a means to dramatically improve the delivery of health care places a heavy burden on the shoulders of those who would actively impede ES and NT technology. The emphasis on the moral error of sin by omission is widely reflected in Western tradition traceable to Biblical tradition. In Matthew chapter 25 we are told of the parable of the master who leaves talents of gold with his servants. One servant, for fear of making a mistake with what was given him, buries the talent in the ground. This servant, labeled "wicked and slothful" in the Bible, reminds us that simple inaction, when we have been given a valuable asset, is not just a lack of doing good, but is in reality evil. There are times that it is not better to be safe than sorry.

Historically, the United States has a proud history of leading the free world in the bold exploration of new technologies. We did not hesitate to apply our best minds in an effort to allow a man to touch the moon. We were not paralyzed by the fear that like the tower of Babel, we were reaching for the heavens. But a far greater challenge stands before us. We have been given two talents of gold. The first, the human embryonic stem cell, the second, nuclear transfer technology. Shall we, like the good steward, take these gifts to mankind and courageously use them to the best of our abilities to alleviate the suffering of our fellow human being, or will we fail most miserably and bury these gifts in the earth? This truly is a matter of life and death. I urge you to stand courageously in favor of existing human life. The alternative is to inherit the wind.

TESTIMONY

U.S. Senate

Senator Bill Frist

In my work as a heart and lung transplant surgeon, I have for years wrestled with decisions involving life, death, health and healing. Having taken part in hundreds of organ and tissue transplants, I've experienced the ethical challenges involved in end-of-life care on numerous occasions. I've seen families faced with the most difficult decision of saying farewell to a loved one. Yet I have also seen their selfless acts in the midst of this sadness to consent to donate living organs and tissues of their loved ones to the benefit of others. Like organ donation, stem cell research forces us to make difficult decisions. While holding great potential to save lives, it also raises difficult moral and ethical considerations.

I am pro-life. My voting record in the Senate has consistently reflected my pro-life philosophy. As a physician my sole purpose has been to preserve and improve the quality of life. The issue of whether or not to use stem cells for medical research involves deeply held moral, religious and ethical beliefs as well as scientific and medical considerations. After grappling with the issue scientifically, ethically and morally, I conclude that both embryonic and adult stem cell research should be federally funded within a carefully regulated, fully transparent framework. This framework must ensure the highest level of respect for the moral significance of the human embryo. Because of the unique interaction between this potentially powerful new research and the moral considerations of life, we must ensure a strong, comprehensive,

Senator Bill Frist (R) is a member of the U.S. Senate from Tennessee. Before being elected to the Senate in 1994 and reelected in 2000, he was one of the nation's leading heart and lung surgeons. Selections from testimony before the U.S. Senate Committee on Appropriations, Subcommittee on Labor, Health, and Human Services, July 18, 2001.

publicly accountable oversight structure that is responsive on an ongoing basis to moral, ethical and scientific considerations.

Embryonic stem cell research is a promising and important line of inquiry. I'm fully aware and supportive of the advances being made each day using adult stem cells. It is clear, however, that research using the more versatile embryonic stem cells has greater potential than research limited to adult stem cells and can, under the proper conditions, be conducted ethically. The prudent course for us as policymakers is to provide for the pursuit of both lines of research—allowing researchers in each field to build on the progress of the other.

***

The following 10 points are essential components of a comprehensive framework that allows stem cell research to progress in a manner respectful of both the moral significance of human embryos and the potential of stem cell research to improve health.

1. **Ban Embryo Creation for Research:** The creation of human embryos solely for research purposes should be strictly prohibited.
2. **Continue Funding Ban on Derivation:** Strengthen and codify the current ban on federal funding for the derivation of embryonic stem cells.
3. **Ban Human Cloning:** Prohibit all human cloning to prevent the creation and exploitation of life for research purposes.
4. **Increase Adult Stem Cell Research Funding:** Increase federal funding for research on adult stem cells to ensure the pursuit of all promising areas of stem cell research.
5. **Provide Funding for Embryonic Stem Cell Research Only from Blastocysts That Would Otherwise Be Discarded:** Allow federal funding for research using only those embryonic stem cells derived from blastocysts that are left over after in vitro fertilization (IVF) and would otherwise be discarded.
6. **Require a Rigorous Informed Consent Process:** To ensure that blastocysts used for stem cell research are only those that would otherwise be discarded, require a comprehensive informed consent process establishing a clear separation between potential donors' primary decision to donate blastocysts for adoption or to discard blastocysts and their subsequent option to donate blastocysts for research purposes. Such a process, modeled in part on well-established and broadly ac-

cepted organ and tissue donation practices, will ensure that donors are fully informed of all their options.

7. **Limit Number of Stem Cell Lines:** Restrict federally-funded research using embryonic stem cells derived from blastocysts to a limited number of cell lines. In addition, authorize federal funding for embryonic stem cell research for five years to ensure ongoing Congressional oversight.
8. **Establish a Strong Public Research Oversight System:** Establish appropriate public oversight mechanisms, including a national research registry, to ensure the transparent, in-depth monitoring of federally-funded and federally-regulated stem cell research and to promote ethical, high quality research standards.
9. **Require Ongoing, Independent Scientific and Ethical Review:** Establish an ongoing scientific review of stem cell research by the Institute of Medicine (IOM) and create an independent Presidential advisory panel to monitor evolving bioethical issues in the area of stem cell research. In addition, require the Secretary of Health and Human Services to report to Congress annually on the status of federal grants for stem cell research, the number of stem cell lines created, the results of stem cell research, the number of grant applications received and awarded, and the amount of federal funding provided.
10. **Strengthen and Harmonize Fetal Tissue Research Restrictions:** Because stem cell research would be subject to new, stringent federal requirements, ensure that informed consent and oversight regulations applicable to federally-funded fetal tissue research are consistent with these new rules.

TESTIMONY

U.S. Senate

Senator Sam Brownback

The central question to this debate remains: Is the young human a life or mere property to be discarded as a master chooses? Destructive embryo research—research which requires the destruction of living human embryos—is deeply immoral, illegal and unnecessary.

I would like to applaud President Bush for his bold and principled stand in defense of the most vulnerable human life. In a letter to the Culture of Life Foundation on May 18, 2001, the President states, "I oppose federal funding for stem cell research that involves destroying living human embryos. I support innovative medical research on life-threatening and debilitating diseases, including promising research on stem cells from adult tissue."

Also, let me state that my good friend, who is testifying alongside me today, Senator Orrin Hatch, and I both share a deep commitment to finding cures for the myriad diseases which plague humanity. I strongly support a doubling in the funding for the National Institutes of Health, as well as other important areas of research. However, there is an area of serious disagreement between us.

Our disagreement lies in the area of destructive embryo research. As you know, the research in question relies on the destruction of human embryos for their so-called, "stem cells." I have opposed the use of human embryos—very early human beings—for this research because the proposed research would result in their death.

Senator Sam Brownback (R) is a member of the U.S. Senate from Kansas. He was elected to the Senate in a special election in 1996 and reelected in 1998. He is the sponsor of the Human Cloning Prohibition Act of 2001 and a leading opponent of embryonic stem cell research. Selections from testimony before the U.S. Senate Committee on Appropriations, Subcommittee on Labor, Health, and Human Services, July 18, 2001.

Unfortunately, over the past few years, there has been a lot of misinformation and confusion about stem cells—as well as the sources of those cells.

In fact, far from lagging behind embryonic stem cells, adult stem cells are already being used in human patients to assist recovery from cancer and leukemia, restore sight to the blind, cure severe combined immune deficiency, and repair damaged bone and cartilage. On the contrary, recent animal trials using embryonic stem cells have shown a disturbing tendency for these cells to form uncontrolled tumors when transplanted. To my knowledge, and my contact with the scientific literature confirms this, there is no embryonic stem cell work even close to treating humans.

Research using adult stem cells, umbilical cord blood, and other approaches are showing much more promising results; in fact, in many cases they are already being used in clinical applications. Recent findings attest to the pluripotent nature of adult stem cells—a finding that renders the so-called "need" for destructive embryo research moot. There are promising avenues of research that do not involve the destruction of innocent human life; and it is those areas which we ought to aggressively fund with taxpayer dollars.

I would also like to point out that many of the proponents of destructive embryo research are now advocating the use of so-called "therapeutic cloning." In fact, at a recent hearing I held on the issue of human cloning both the President of the Biotechnology Industry Organization (BIO) and a representative from the American Society for Cell Biology emphasized their strong support for so-called "therapeutic cloning" as the ultimate source of embryonic stem cells that won't face "rejection."

Dr. Rudolph Janeisch testified that with "therapeutic cloning," "no rejection will occur, because these cells, which come from the cloned embryonic stem cells, are of the same immunological makeup as the patient's cells itself."

The testimony of both Dr. Janeisch and Mr. [Carl] Feldbaum recognize that for the purposes of possible clinical applications, particularly to avoid possible tissue rejection, human cloning is the logical next step—or so-called "therapeutic cloning." This means that live embryos created by researchers can be experimented on and killed at the leisure of researchers for the purported benefit of the patient.

The slippery slope of embryonic stem cell research, which is in itself immoral, ultimately leads to cloning and both must be stopped. I say slippery slope, because the current proposals seek to undo any principled limitation by rejecting true principle. The principle being denied in this case is the dignity of the young human, effectively making the human embryo equal to mere plant or animal life, or property, to be disposed of according to human choice governed by mere legal and pragmatic considerations.

Those who advocate destructive embryo research have begun changing the terms in the debate, one would almost suspect in a blatant attempt to confuse the American people. In fact, when the Coalition for the Advancement of Medical Research conducted a poll earlier this year they claimed that 70 percent of Americans "supported research using stem cells from excess fertilized eggs."

Compare it with the question posed by International Communications Research (who performs polling for ABC news), which found that 70 percent of Americans opposed federal funding for experiments which would destroy "live human embryos" for their stem cells.

Regardless of polls which can be manipulated and public sentiment which can change, the fact remains that we simply do not need to do any research which relies on the destruction of human beings. I would argue that we should not do the research for moral reasons; but even from a pragmatic point of view the research is simply unnecessary.

There are many Americans who share this view. In fact, I have received letters from several people who are suffering from a disease, and who will never accept treatment that has been provided by the killing of another human being.

The coercive power of the federal government should not be used to force taxpayers to pay for the destruction of innocent human beings. There are countless people who have deep moral objections to this and forcing them to pay for the destruction of human lives would be a grave injustice to them and to the values they hold.

IT'S A BIRD, IT'S A PLANE, IT'S . . . SUPERCLONE?

Katha Pollitt and Critics

Is human cloning a feminist issue? Two cloning bans are currently winding their way through Congress: In the Senate, the Human Cloning Prohibition Act seeks to ban all cloning of human cells, while a House version leaves a window open for cloning stem cells but bans attempts to create a cloned human being. Since both bills are the brain-children of antichoice Republican yahoos, who have done nothing for women's health or rights in their entire lives, I was surprised to get an e-mail inviting me to sign a petition supporting the total ban, organized by feminist heroine Judy Norsigian of the Boston Women's Health Book Collective (the producers of *Our Bodies, Ourselves*) and signed by Ruth Hubbard, Barbara Seaman, Naomi Klein and many others. Are feminists so worried about "creating a duplicate human" that they would ban potentially useful medical research? Isn't that the mirror image of antichoice attempts to block research using stem cells from embryos created during in vitro fertilization?

My antennae go up when people start talking about threats to "human individuality and dignity"—that's a harrumph, not an argument. The petition raises one real ethical issue, however, that hasn't gotten much attention but by itself justifies a ban on trying to clone a person: The necessary experimentation—implanting clonal embryos in surrogate mothers until one survives till birth—would involve serious medical risks for the women and lots of severely defective babies. Dolly, the cloned Scottish sheep, was the outcome of a process that included hundreds of monstrous discards, and Dolly

Katha Pollitt is a columnist for *The Nation* and author of *Subject to Debate: Sense and Dissents on Women, Politics, and Culture.* This article appeared in *The Nation* on July 23/30, 2001. The letters to the editor were published in *The Nation* on October 8, 2001.

herself has encountered developmental problems. That's good reason to go slow on human research—especially when you consider that the people pushing it most aggressively are the Raelians, the UFO-worshiping cult of techno-geeks who have enlisted the services of Panayiotis Zanos, a self-described "cowboy" of assisted reproduction who has been fired from two academic jobs for financial and other shenanigans.

Experimental ethics aside, though, I have a hard time taking cloning seriously as a threat to women or anyone else—the scenarios are so nutty. Jean Bethke Elshtain, who took a break from bashing gay marriage to testify last month before Congress against cloning, wrote a piece in *The New Republic* in 1997 in which she seemed to think cloning an adult cell would produce another adult—a carbon of yourself that could be kept for spare parts, or maybe a small army of Mozart xeroxes, all wearing knee breeches and playing the *Marriage of Figaro.* Actually, Mozart's clone would be less like him than identical twins are like each other: He would have different mitochondrial DNA and a different prenatal environment, not to mention a childhood in twenty-first-century America with the Smith family rather than in eighteenth-century Austria under the thumb of the redoubtable Leopold Mozart. The clone might be musical, or he might be a billiard-playing lounge lizard, but he couldn't compose *Figaro.* Someone already did that.

People thinking about cloning tend to imagine *Brave New World* dystopias in which genetic engineering reinforces inequality. But why, for example, would a corporation go to the trouble of cloning cheap labor? We have Mexico and Central America right next door! As for cloning geniuses to create superbabies, good luck. The last thing most Americans want are kids smarter than they are, rolling their eyeballs every time Dad starts in on the gays and slouching off to their rooms to I-M [Internet Message] other genius kids in Sanskrit. Over nine years, only 229 babies were born to women using the sperm bank stocked with Nobel Prize winners' semen—a tiny fraction, I'll bet, of those conceived in motel rooms with reproductive assistance from Dr. Jack Daniel's.

Similarly, cloning raises fears of do-it-yourself eugenics—designer babies "enhanced" through gene manipulation. It's hard to see that catching on, either. Half of all pregnancies are unintended in this country. People could be planning for "perfect" babies today—preparing for conception by giving up cigarettes and alcohol and unhealthy foods, reading Stendhal to their fetuses in French. Only a handful of yuppie control freaks actually do this, the same ones who obsess about getting their child into a nursery school that leads straight to Harvard. Those people are already the "genetic elite"—white, with lots of family money. What do they need genetic enhancement for? They think they're perfect now.

Advocates of genetic tinkering make a lot of assumptions that opponents tacitly accept: for instance, that intelligence, talent and other qualities are genetic, and in a simple way. Gays, for example, worry that discovery of a "gay gene" will permit selective abortion of homosexual fetuses, but it's obvious that same-sex desire is more complicated than a single gene. Think of Ancient Greece, or Smith College. Even if genetic enhancement isn't the pipe dream I suspect it is, feminists should be the first to understand how socially mediated supposedly inborn qualities are—after all, women are always being told anatomy is their destiny.

There's a strain of feminism that comes out of the women's health movement of the seventies that is deeply suspicious of reproductive technology. In this view, prenatal testing, in vitro fertilization and other innovations commodify women's bodies, are subtly coercive and increase women's anxieties, while moving us steadily away from experiencing pregnancy and childbirth as normal, natural processes. There's some truth to that, but what about the side of feminism that wants to open up new possibilities for women? Reproductive technology lets women have children, and healthy children, later; have kids with lesbian partners; have kids despite disabilities and illness. Cloning sounds a little weird, but so did in vitro in 1978, when Louise Brown became the first "test tube baby." Of course, these technologies have evolved in the context of for-profit medicine; of course they represent skewed priorities, given that 43 million Americans lack health insurance and millions worldwide die of curable diseases like malaria. Who could argue that the money and brain power devoted to cloning stem cells could not be better used on something else? But the same can be said of every aspect of American life. The enemy isn't the research, it's capitalism.

LETTERS TO THE EDITOR

Katha Pollitt is my favorite *Nation* columnist, but guess what, Katha, you've got my objections to cloning embryo stem cells all wrong.

Maybe now that it's public knowledge that researchers have been buying women's eggs so they can make human embryos for research, you may be getting the point. But if not, let me explain. Human eggs don't grow on trees. They are embedded deep inside women's bodies; not easy to get at, like sperm. To collect more than one at a time means you first have to give women hormones to shut their ovaries down. You then have to hyperstimulate their ovaries with hormones of another sort so that many more than the customary single monthly follicle and its egg mature. At the right time, you then puncture each follicle and suck out its egg. Sound good for women? Steptoe and Edwards, the scientific "fathers" of Louise Brown, didn't give her mother hormones because they feared it wasn't safe. They waited patiently for a follicle to mature and then collected the egg that eventually became Louise. But the fertility industry doesn't have time for such niceties. Now it's hormones and mass production.

The concerns Pollitt imputes to me are not what worries me. I worry about what this new "need" for human embryos will do to women. And do you know what? We may never know the answer, because in countries with proper healthcare systems where proper health records are kept, people are not permitted to buy and sell body parts. "The enemy isn't the research," Pollitt writes, "it's capitalism." Wrong again: The enemy is *research under capitalism*.

RUTH HUBBARD
Professor emeritus of biology
Harvard University
Board member, Council for Responsible Genetics

It's unfortunate that the usually perceptive Katha Pollitt completely misses the point about human cloning in her column on this subject. Kids produced by in vitro fertilization are one thing: They are made from the standard starting materials—an egg and a sperm. The donor of either may request anonymity, but the resulting infant is guaranteed to be a full-fledged member of the human species, biologically speaking (as well as socially and legally, although these connections are rapidly being eroded in the current environment—see Lori Andrews, *The Clone Age,* Henry Holt, 1999).

A clone is quite a different animal, however. It is constructed of parts of cells (an egg missing its nucleus; the nucleus of an adult cell) that never meet in the course of reproduction. Evolution has never had to deal with, and arrive at correctives for, the errors introduced into the developmental process resulting from this atypical combination of cell parts. No wonder virtually all attempts at animal cloning have led to fetal deaths, multiple birth defects or severe health problems later in the lives of even the most sound-looking clones. This is not a set of problems that can be worked out in mice before confidently being attempted in humans; it is probably too complex to be fully controlled, and in any case, each species presents unique complications.

A Massachusetts company, Advanced Cell Technologies, has announced that it is now producing clonal human embryos as a first step in producing donor-matched therapeutic stem cells. And now biotechnology industry representatives have begun to make common cause with some of their anti-choice beneficiaries in Congress in trying to define such embryos as "not true human embryos" in order to thwart laws against their production and manipulation. Indeed, if Pollitt's blasé attitude toward the production of full-term human clones becomes prevalent, we can look forward to the day when the not-quite-natural, not-quite-artificial products of human cloning experiments (disconnected, as they would be, from any social network other than that defined by ownership rights) are also redefined as "not true humans." This would open the way to their finding use as sources of transplantable organs, experimental laboratory models or perhaps, for the most presentable examples, wounded hero status in the march of reproductive technology. Would Pollitt flip off concerns about "threats to 'human individuality and dignity'" in this not very distant brave new world?

STUART A. NEWMAN
Professor of cell biology,
New York Medical College
Board member, Council for Responsible Genetics

On the question of human cloning, Katha Pollitt's usually reliable political insight has failed her. She dismisses the pro-choice statement calling for bans on human cloning—signed to date by more than a hundred women's health and reproductive rights leaders—on the grounds that the pending Congressional bills to prohibit cloning are the "brainchildren of anti-choice Republican yahoos."

But that's precisely the point: Human cloning and genetic manipulation are feminist-liberal-progressive-radical issues. We leave them to the anti-choice crowd at our considerable peril.

The recent deluge of news about stem cells has generated a great deal of confusion about cloning. Two clarifications are key: First, opposition to cloning can and does co-exist with support for research on embryonic stem cells, using embryos from in vitro fertilization procedures. Stem cell research and embryo cloning intersect, but they are technically distinguishable—and vastly different politically.

Second, looking at human cloning through the lens of abortion politics blurs and distorts its meaning. The prospect of cloned or genetically "enhanced" children is ominous because it could so easily trigger an unprecedented kind of eugenics, one implemented not by state coercion but by upscale marketing campaigns for designer babies.

Pollitt thinks this scenario unlikely. I invite her to reconsider. The marginal figures she mentions—the Raelians and the cowboy fertility doctor Panos Zavos—are not the only champions of human cloning, and they are far from the most dangerous.

Already biotech companies are jockeying for patents on procedures to clone and manipulate human embryos. And for several years now, a disturbing number of influential scientists, biotech entrepreneurs, bioethicists and others have been actively promoting human cloning and genetic redesign. Some are open about their ambition to set humanity on a eugenic path and to "seize control of human evolution."

One example among many is Princeton University molecular biologist Lee Silver. In multiple appearances on national television and in the newsweeklies, Silver has plugged the "inevitable" emergence of a genetic caste system in which the "GenRich" rule and the "Naturals" work as "low-paid service providers." Like others of his persuasion, he seems quite ready to abandon any pretense of commitment to equality—or even to a common humanity.

Pollitt is right to caution against accepting wildly overblown claims about the power of genes to determine everything from sexual orientation to homelessness. But it would be foolish to overlook the rapidly expanding powers of genetic manipulation, or to dismiss the possibility that the advocates of a "post-human" future will achieve enough mastery over the human genome to wreak enormous damage—biologically, culturally and politically.

Free-market eugenics is not science fiction or far off. It is an active political agenda that must be urgently opposed.

MARCY DARNOVSKY
Exploratory Initiative on the New Human Genetic Technologies

KATHA POLLITT REPLIES

"It's a Bird, It's a Plane, It's . . . Superclone?"— my column on cloning—was probably the most unpopular "Subject to Debate" ever. Clearly this is a vexed subject, with many aspects, some of which are noted in the letters above.

The hormone-stimulated ripening and extraction of eggs, which Ruth Hubbard vividly describes, are, as she notes, the basis of much assisted reproduction, including now-routine procedures like in vitro fertilization with one's own eggs. Indeed, many college newspapers advertise for egg donors, and many students are willing to go through the extraction process and to take on its risks in return for substantial fees. Cloning would expand this market—how much, we don't know—but the market already flourishes.

While I, too, am troubled by so many women undergoing procedures whose long-term safety is still unknown—I mentioned this in my column as a fair objection to cloning—the fact is that every day all sorts of people take risks for money, or knowledge, or pleasure, or survival. What makes eggs so sacred? And would Hubbard approve of cloning embryos if the eggs were obtained in the "patient," old-fashioned way?

Stuart Newman and Marcy Darnovsky raise "brave new world" scenarios that to me do indeed sound farfetched and wild, and not even all bad—why would it be bad to "design" healthy babies, cloned or not? In any case, cloning seems like an odd place to begin worrying about a society divided into classes destined from birth for different levels of health, wealth and personal development: We live in that society now!

A NIGHTMARE OF A BILL

Charles Krauthammer

Hadn't we all agreed—we supporters of stem cell research—that it was morally okay to destroy a tiny human embryo for its possibly curative stem cells because these embryos from fertility clinics were going to be discarded anyway? Hadn't we also agreed that human embryos should not be created solely for the purpose of being dismembered and then destroyed for the benefit of others?

Indeed, when Senator Bill Frist made that brilliant presentation on the floor of the Senate supporting stem cell research, he included among his conditions a total ban on creating human embryos just to be stem cell farms. Why, then, are so many stem cell supporters in Congress lining up behind a supposedly "anti-cloning bill" that would, in fact, legalize the creation of cloned human embryos *solely* for purposes of research and destruction?

Sound surreal? It is.

There are two bills in Congress regarding cloning. The Weldon bill bans the creation of cloned human embryos for any purpose, whether for growing them into cloned human children or for using them for research or for their parts and then destroying them.

The competing Greenwood "Cloning Prohibition Act of 2001" prohibits only the creation of a cloned child. It protects and indeed codifies the creation of cloned human *embryos* for industrial and research purposes.

Under Greenwood, points out the distinguished bioethicist Leon Kass, "embryo production is explicitly licensed and treated like drug manufacture." It becomes an industry, complete with industrial secrecy protections.

This essay was originally published in the *Washington Post* on July 27, 2001.

Greenwood, he says correctly, should really be called the "Human Embryo Cloning Registration and Industry Facilitation and Protection Act of 2001."

Greenwood is a nightmare and an abomination. First of all, once the industry of cloning human embryos has begun and thousands are being created, grown, bought and sold, who is going to prevent them from being implanted in a woman and developed into a cloned child?

Even more perversely, when that inevitably occurs, what is the federal government going to do: Force that woman to abort the clone?

Greenwood sanctions, licenses and protects the launching of the most ghoulish and dangerous enterprise in modern scientific history: the creation of nascent cloned human life for the sole purpose of its exploitation and destruction.

What does one say to stem cell opponents? They warned about the slippery slope. They said: Once you start using discarded embryos, the next step is creating embryos for their parts. Senator Frist and I and others have argued: No, we can draw the line.

Why should anyone believe us? Even before the president has decided on federal support for stem cell research, we find stem cell supporters and their biotech industry allies trying to pass a bill that would cross that line—not in some slippery-slope future, but right now.

Apologists for Greenwood will say: Science will march on anyway. Human cloning will be performed. Might as well give in and just regulate it, because a full ban will fail in any event.

Wrong. Very wrong. Why? Simple: You're a brilliant young scientist graduating from medical school. You have a glowing future in biotechnology, where peer recognition, publications, honors, financial rewards, maybe even a Nobel Prize await you. Where are you going to spend your life? Working on an outlawed procedure? If cloning is outlawed, will you devote yourself to research that cannot see the light of day, that will leave you ostracized and working in shadow, that will render you liable to arrest, prosecution and disgrace?

True, some will make that choice. Every generation has its Kevorkian. But they will be very small in number. And like Kevorkian, they will not be very bright.

The movies have it wrong. The mad scientist is no genius. Dr. Frankensteins invariably produce lousy science. What is Kevorkian's great contribution to science? A suicide machine that your average Hitler Youth could have turned out as a summer camp project.

Of course you cannot stop cloning completely. But make it illegal and you will have robbed it of its most important resource: great young minds. If we act now by passing Weldon, we can retard this monstrosity by decades. Enough time to regain our moral equilibrium—and the recognition that the human embryo, cloned or not, is not to be created for the sole purpose of being poked and prodded, strip-mined for parts and then destroyed.

If Weldon is stopped, the game is up. If Congress cannot pass the Weldon ban on cloning, then stem cell research itself must not be supported either—because then all the vaunted promises about not permitting the creation of human embryos solely for their exploitation and destruction will have been *shown in advance* to be a fraud.

STEM-CELL RESEARCH

DON'T DESTROY HUMAN LIFE

Robert P. George

The debate over "harvesting" stem cells from human embryos has forced policy makers to think about a question that they would rather avoid: When does a new human being begin? Of course, this is a question that policy makers should have been thinking about in the context of the debate over abortion. They were relieved of that task by the Supreme Court's misbegotten decision to legalize abortion by judicial fiat. This time no court will let them off the hook.

Still, they are wriggling. White House advisers and members of Congress are looking for a solution or compromise that will enable them to avoid the question whether the destruction of human embryos is in fact the killing of human beings.

They won't find one.

The most recent attempt is by William Frist—the highly respected prolife Republican senator from Tennessee who happens also to be an eminent physician. The Frist proposal would ban the funding of research involving the creation of embryos for stem cell harvesting, while permitting the harvesting of stem cells from "excess" embryos created by in vitro fertilization. The argument is: If they will be disposed of anyway, why not make good use of them by dismembering them and obtaining their stem cells?

Robert P. George is director of the James Madison Program in American Ideals and Institutions and McCormick Professor of Jurisprudence at Princeton University. His many essays and books include *Making Men Moral: Civil Liberties and Public Morality*, *In Defense of Natural Law,* and *The Clash of Orthodoxies*. This essay originally appeared in the *Wall Street Journal* on July 30, 2001.

The trouble with the proposal is that it assumes, on the one hand, that embryos are human beings and therefore should not be brought into being for purposes of destructive research—no matter how great the possible scientific and medical benefits. Yet it allows the destruction of some embryos.

We would, I hope, never permit or fund the harvesting of organs from retarded human infants, demented or terminally ill patients, or even death row prisoners. It wouldn't matter that death was expected in five months or five minutes. Nor would it matter that a dying patient, for example, was unconscious, even permanently unconscious as a result of a coma. Nor would we factor into our deliberation any consideration of the promise of what science or medicine could do with the organs. We wouldn't tolerate killing for purposes of harvesting body parts because it is inconsistent with the inherent dignity of all human beings.

So there is no avoiding the question: Are embryos, or are they not, human beings?

What is a human being? He or she is a whole, living member of the species *Homo sapiens*. Plainly gametes (sperm cells and ova) are not human beings. They are parts of other human beings. They lack the epigenetic primordia for internally directed growth and maturation as a distinct, complete, self-integrating, human organism. The same is true of somatic cells (such as skin cells).

Modern science shows that human embryos, by contrast, are whole, living members of the human species, who are capable of directing from within their own integral organic functioning and development into and through the fetal, infant, child, and adolescent stages of life and ultimately into adulthood.

It is not that a human embryo merely has the potential to "become a life" or "become a human being." He or she (for sex is determined at the beginning of life) is already a living human being. In this crucial respect, the embryo is like the fetus, infant, child, and adolescent. The being that is now you or me is the same being that was once an adolescent, and before that a toddler, and before that an infant, and before that a fetus, and before that an embryo. To have destroyed the being that is you or me at any of these stages would have been to destroy you or me.

In the current debate, the question whether a human embryo is a human being is usually ignored or evaded. When it has been faced, the arguments advanced for denying that embryos are human beings have been astonishingly weak. (Understandably, proponents of destructive embryo research have tried to shift the focus to its potential benefits.)

Some commentators say that human embryos don't "look like" human beings. The answer is that they look exactly like the human beings they are, that is, human beings in the embryonic stage of their existence. Others try to make something of the fact that embryos are tiny, or very immature, or dependent for full development upon implantation. Sen. Orrin Hatch has gone so far as to make the location of an embryo—in a dish or refrigeration unit rather than in a mother's womb—determinative of its moral status. But anybody who gives the matter some thought should recoil from the idea that factors such as size, stage of development, location, and state of dependency can be a basis for denying rights to human beings.

Then there is the claim that the argument for the human status of the early embryo depends on controversial religious premises about "ensoulment." It does not. The question is not about embryos' eternal destiny. That is a religious matter. (One on which the Catholic Church, by the way, has no official position.) There is no need for those of us who oppose embryo destruction to appeal to religion. The science will do just fine. We would be very pleased if those on the other side would agree that the scientific facts about when new human beings begin should determine whether government should fund research requiring their deliberate destruction.

Of course some proponents of stem-cell research are willing to concede the embryos are human beings. My Princeton colleague Peter Singer and other outright utilitarians deny that there is a principle of inherent human dignity that stands as an absolute bar to killing some people for the sake of a putative "greater good." So they typically see no moral reason not to dismember living embryos for their stem cells. By the same token, they see no moral reason not to kill human beings at any stage of maturity when, as they suppose can happen, some calculus of utility tips the scales in that direction. Hence, Mr. Singer's notorious defense of infanticide of handicapped newborns.

The concept of the "human non-person"—a human being whose life can be deliberately destroyed, or who can be mutilated or enslaved, to serve the interests of others—richly deserves the ignominy in which it has come to be held. Let us not accept the devil's bargain of reviving [it] in the mere hope of scientific advances.

SPEECH

U.S. House of Representatives

Representative James Greenwood

Mr. Speaker, this is a matter of values. It is a matter of how much one values our ability to end human suffering and to cure disease.

No one in this House should be so arrogant as to assume that they have a monopoly on values, that their side of an argument is the values side and the other's is not. This is a matter of how much we value saving little children's lives and saving our parents' lives.

There has been talk on the floor about creating embryo factories. Most of that talk I think has been conducted by people who do not understand the first thing about this research.

Here is how one could create an embryo factory. We would get a long line of women who line up in a laboratory and say, would you please put me through the extraordinarily painful process of superovulation because I would like to donate my eggs to science.

Does anybody think that is going to happen? Of course it is not going to happen. We are going to take this research, and this research involves a very small handful of cells. In the natural world, every day millions of cells, millions of eggs, are fertilized, and they do not adhere to the wall of the uterus. They are flushed away. That is how God does God's work.

Representative James Greenwood (R) is a member of the House of Representatives from Pennsylvania. He was elected in 1992 and was named Biotechnology Industry Organization Legislator of the Year in 1998. He was the sponsor of unsuccessful legislation in 2001 to ban only "reproductive cloning" and allow "therapeutic cloning." Included here are selections from two of his speeches during the floor debate on human cloning, *Congressional Record*, July 31, 2001.

In in vitro fertilization clinics, every day thousands of eggs are fertilized, and most of them are discarded. That is the way loving parents build families who cannot do it otherwise. No one is here to object to that. Thousands of embryos are destroyed.

We are talking about a handful, a tiny handful of eggs that are utilized strictly for the purpose of understanding how cells transform themselves from somatic to stem and back to somatic, because when we understand that, we will not need any more embryonic material. We will not need any cloned eggs. We will have discovered the proteins and the growth factors that let us take the DNA of our own bodies to cure that which tortures us.

That is the value that I am here to stand for, because I care about those children, and I care about those parents, and I care about those loved ones who are suffering.

I am not prepared as a politician to stand on the floor of the House and say, I have a philosophical reason, probably stemmed in my religion, that makes me say, you cannot go there, science, because it violates my religious belief.

I think it violates the constitution to take that position.

And on the question of whether or not we can do stem cell research with the Weldon bill in place, I would quote the American Association of Medical Colleges. It says, "H.R. 2505 would have a chilling effect on vital areas of research that could prove to be of enormous public benefit." The Weldon bill would be responsible for having that chilling effect on research.

The Greenwood substitute stops reproductive cloning in its tracks, as it ought to be stopped, but allows the research to continue, and I would advocate its support.

***

We have had a good two hours of debate, and it has been encouraging to see the extent to which Members of Congress have been able to grapple with this very complicated issue.

Unfortunately, the Members who are speaking are the ones who have mastered it. We will have a vote within the hour and unfortunately most Members will come here pretty confused about the issue.

Let me try to simplify the issue once again and ask that we try to avoid some of the *ad hominem* argument that I think is beginning, and the hostility, frankly, that is beginning to develop on the floor on this issue. This is not a question about who has values and who stands for human life and who does not. It is a very legitimate and important and historic debate about how it is that we are able to use the DNA that God put into our own bodies, use the brain that God gave us to think creatively, and to employ this research to save the lives of men, women and children in this country and throughout the world and to rescue them from terribly debilitating and life-shortening diseases.

SPEECH

U.S. House of Representatives

Representative Dennis J. Kucinich

Mr. Speaker, the pro-life/pro-choice debate has centered on a disagreement about the rights of the mother and whether her fetus has legally recognized rights. But in this debate on human cloning, there is no woman. The reproduction and gestation of the human embryo takes place in the factory or laboratory; it does not take place in a woman's uterus.

Therefore, the concern for the protection of a woman's right does not arise in this debate on human cloning. There is no woman in this debate. There is no mother. There is no father. But there is a corporation functioning as creator, investor, manufacturer, and marketer of cloned human embryos. To the corporation, it is just another product with commercial value. This reduces the embryo to just another input.

What we are discussing today in the Greenwood bill is the right of a corporation to create human embryos for the marketplace, and perhaps they will be used for research, perhaps they will be just for profit, all taking place in a private lab.

But is this purely a private matter, this business of enucleating an egg and inserting DNA material from a donor cell, creating human embryos for research, for experimentation, for destruction, or perhaps, though not intended, for implantation? Is this just a matter between the clone and the corporation, or does society have a stake in this debate?

We are not talking about replicating skin cells for grafting purposes. We are not talking about replicating liver cells for transplants. We are talking

Representative Dennis J. Kucinich (D) is a member of the House of Representatives from Ohio. He was elected in 1996 and is chairman of the Congressional Progressive Caucus. Included here is his speech during the floor debate on human cloning, *Congressional Record*, July 31, 2001.

about cloning whole embryos. The industry recognizes there is commercial value to the human life potential of an embryo, but does a human embryo have only commercial value? That is the philosophical and legal question we are deciding here today.

The Greenwood bill, which grants a superior cloning status to corporations, would have us believe that human embryos are products, the inputs of mechanization, like milling timber to create paper, or melting iron to create steel, or drilling oil to create gasoline. Are we ready to concede that human embryos are commercial products? Are we ready to license industry so it can proceed with the manufacture of human embryos?

If this debate is about banning human cloning, we should not consider bills which do the opposite. The Greenwood substitute to ban cloning is really a bill to begin to license corporations to begin cloning. Though the substitute claims to be a ban on reproductive cloning, it makes this nearly impossible by creating a system for the manufacture of cloned embryos. It does not have a system for Federal oversight of what is produced and does not allow for public oversight. The substitute allows companies to proceed with controversial cloning with nearly complete confidentiality.

Cloning is not an issue for the profit-motivated biotech industry to charge ahead with; cloning is an issue for Congress to consider carefully, openly, and thoughtfully. That is why I support the Weldon bill. I urge that all others support it as well.

CLONING, STEM CELLS, AND BEYOND

Eric Cohen and William Kristol

Last week's vote in the House to ban human cloning is something to celebrate. It may even be something momentous. The House passed, by 265 to 162, a bill sponsored by representative David Weldon of Florida that would ban the creation of all human clones. It rejected an alternative sponsored by Pennsylvania representative James Greenwood, and backed by the biotech lobby, that would have allowed the creation of cloned human embryos to be used for medical research and then destroyed.

The Greenwood forces had corporate money and much of enlightened opinion behind them. They overpromised, misled, and demagogued, claiming, for example, that cloned-embryo research could one day "end human suffering," that cloned embryos "are not really embryos at all," and that a vote against such research is a "sentence of death for millions of Americans."

But the majority of the House—a larger majority than expected—refused to listen. They chose instead to halt (or try to halt) what Charles Krauthammer has described as "the most ghoulish and dangerous enterprise in modern scientific history: the creation of nascent cloned human life for the sole purpose of its exploitation and destruction." They defied the wishes of the medical research establishment, the biotech industry, and the health-at-any-cost humanitarians. They drew a bright moral line, which even the most well-meaning scientists would not be permitted to cross.

Whether this line will hold in the long run—and even whether the Senate will pass a similar cloning ban—is an open question. For while last week's House vote struck a blow against a Brave New World, it represents only the

This essay originally appeared in the *Weekly Standard* on August 13, 2001 (printed on August 3, 2001).

first public engagement in what will surely be a prolonged struggle, not just over cloning and stem cells, but over whether and how to regulate, control, and shape the genetic revolution that is upon us.

One lesson of last week's debate is that everyone claims to be horrified by the prospect of live human clones. Even the Greenwood bill ostensibly banned reproductive cloning. This suggests a broad willingness to accept some moral limitations, enforced in law, on scientific "progress." It suggests we still believe there are great and obvious evils that no amount of utilitarian or libertarian reasoning can justify, and which we must regulate, forbid, and criminalize in the public interest.

But we have also learned something else: Over one-third of the House of Representatives believes that corporations and researchers—like Advanced Cell Technology in Worcester, Massachusetts, which has already begun a research cloning project—should be left alone in the hope that cloned-embryo farms will one day prove a useful source of embryonic stem cells. And we know that majorities in both the House and Senate support federal funding for embryonic stem cell research, at least when the embryos are "leftovers" from in vitro fertilization clinics. Nor have we seen any urgent effort to ban the creation of embryos by private organizations—like the Jones Institute in Norfolk, Virginia—that pay women to help produce embryos for research and destruction.

And despite all the publicity surrounding the president's pending decision on embryonic stem cells, it is worth noting that his decision will be a limited one, touching only on the question of federal funding of research on stem cell lines derived from "spare" in vitro embryos. Even if the president maintains the current ban on funding, Congress will likely challenge him with a bill of its own—and may well try to broaden the permissible uses of federal funds. And whatever the president and Congress decide about federal funding, this research will presumably proceed apace in the private sector—and not just on leftover in vitro embryos but on embryos created solely for research and destruction.

All of this means that last week's cloning debate in the House and President Bush's imminent stem cell decision are just the tip of the iceberg. The dilemmas over cloning and stem cells will inevitably force a much larger debate about where the modern technological project is heading: Is it moral to harvest potential lives to help existing ones? How about improving potential life through genetic engineering? Isn't the question of how stem cells may be used as morally troubling as the question of how they have been obtained? How reasonable is it, anyway, to try to end all disease and suffering? Do we have the wisdom and the will to preserve a distinction between medical therapy and eugenic enhancement? A line between a better human world and a new inhuman one?

In this opening skirmish—call it "the cloning/stem cell moment"—four basic positions have emerged. Each represents a different set of moral, political, and practical judgments about what is fundamentally desirable and what is not, and about whether even seemingly desirable advances may have very undesirable consequences. We might call the four camps the hubristic scientists, the squishy liberals, the anguished moderates, and the anti-Brave New Worlders.

HUBRISTIC SCIENTISTS

The hubristic scientists favor medical progress at all costs, and are willing to use any means necessary to further unfettered research, which they equate with the good of mankind. To defend this position they deploy a number of strategies, not all of them true or consistent: the claim that mere legislators and uninformed citizens lack the expertise to make decisions about science; the claim that any "metaphysical" arguments for restricting science are unconstitutional transgressions against the separation of church and state; the assertion that because science is limited ("a method, not a faith," as biotech lobbyist Carl Feldbaum put it), religious people should not worry about its excesses; that because human beings are "more than our genetic make-ups," we should allow the geneticists to do whatever they deem necessary with the human genome; that nearly all religious people really want the fruits of the biological enterprise, even if their values initially give them pause; that the spirit of religion and the spirit of science are really the same; and, finally, the insistence that things are not what they seem—or, in this particular debate, that embryos are not embryos and that the Weldon ban on human cloning is really an effort to undermine in vitro fertilization, the right to abortion, and indeed decades and centuries of medical progress.

Greenwood and his allies used all these strategies on the House floor:

"This is Congress again playing scientist," said Louise M. Slaughter, Democrat of New York.

"Now, here we are making a decision like we were the house of cardinals on a religious issue when, in fact, scientists are struggling to find out how human beings actually work," said Jim McDermott, Democrat of Washington.

"I am not prepared as a politician to stand on the floor of the House and say, I have a philosophical belief, probably stemmed in my religion, that makes me say, you cannot go there, science, because it violates my religious belief," said James Greenwood, Republican of Pennsylvania.

And Greenwood again, this time claiming to have God on his side: "It is a very legitimate and important and historic debate about how it is that we are able to use the DNA that God put into our own bodies, use the brain that God gave us to think creatively, and to employ this research to save the lives of men, women, and children in this country and throughout the world and to rescue them from terribly debilitating and life-shortening diseases."

Conspicuous on the House floor was contempt for "theocrats" who would stop the compassionate march of medical progress—together with brazen confidence that God wants science to proceed unregulated. It was altogether an odd mixture of the hubris of the medical researcher seeking to lead his fellow men beyond nature, and the sentimentality of the post-Communist romantic, who sees in genetic science man's new hope for building a kind, just, and liberated heaven on earth. If the House debate is any indication, the path from such hubris and sentimentality to what C.S. Lewis called "the abolition of man" is quick and direct.

SQUISHY LIBERALS

The second position is that of the squishy liberals, best exemplified perhaps by the *Washington Post*. In October 1994 a National Institutes of Health panel of experts recommended that the government fund research that involved creating and destroying human embryos for research purposes alone. The *Post* disagreed, in a sharp editorial that called for "drawing the line." "The creation of human embryos specifically for research that will destroy them is unconscionable," the paper wrote. "The government has no business funding it. . . . It is not necessary to be against abortion rights, or to believe human life literally begins at conception, to be deeply alarmed by the notion of scientists' purposely causing conceptions in a context entirely divorced from even the potential of reproduction."

Fast forward to last week. On the day of the cloning debate, a *Post* editorial entitled "Cloning Overkill" sang a very different tune. All the caution and outrage and commitment to "society's ability to make distinctions" were gone. Now swept up in enthusiasm for stem cell research, the *Post* argued:

"The bill to ban all human cloning, proposed by Rep. David Weldon (R-Fla.), goes well beyond any consensus society has yet reached. . . . At issue is not the withholding of federal funding from research some find morally troubling; rather, the Weldon bill would criminalize the field of cloning entirely. . . . A complete cloning ban could block many possible clinical applications of stem cell research." And the only way those "applications" will be discovered is by creating cloned human embryos for research and destruction—the very thing the paper seven years earlier had deemed "unconscionable."

This is the way of the squishy liberals: They temporarily affirm some moral limits to scientific progress, only to cave when those limits are actually tested by a new wave of medical promises. They are putty in the hands of the less scrupulous avatars of "progress," who use the rhetoric of limits (or pseudo-limits) as a tactic against those who would resist them.

Thus, in the media crusade to win federal funding for embryonic stem cell research, advocates have made their case largely on the grounds that embryos left over from in vitro fertilization will be destroyed anyway. But the House vote shows that many pro-research congressmen are willing to go much further: 178 members (153 of them Democrats) voted to authorize the creation and destruction of cloned embryos.

Here the bait-and-switch dishonesty is remarkable. On July 27, over 200 members of the House wrote President Bush "to express our strong support for federal funding of embryonic stem cell research." The letter continues:

> The reports the week of July 9 that a Virginia laboratory has created human embryos to obtain stem cells for research purposes and a Massachusetts firm aims to create embryos using cloning techniques to derive stem cells for therapeutic purposes, make plain that this research, replete with moral, ethical, and scientific issues, is occurring in the private sector even as the federal government debates the issues. The only way to ensure that embryonic stem cell research is conducted with strict ethical and legal guidelines is to provide federal funding and oversight.

Signing the letter were James Greenwood, Peter Deutsch, and 165 others who voted for the Greenwood bill—the very purpose of which was to authorize the cloning of embryos that this letter pretends to find so alarming.

A vote for the Greenwood bill was a vote for the creation of embryos solely for research and destruction, nothing else. It was a vote for the very thing the *Washington Post*—and many defenders of fetal tissue research in the early 1990s—once explicitly rejected: creation for destruction. And so it is that the alliance of the hubristic scientists and the squishy liberals ensures that some moral limits are no limits at all—just bumps in the road.

ANGUISHED MODERATES

Which raises the question: Can real lines be drawn? Can limits be set and coherent and lasting distinctions made? For example, Republican senator Bill Frist of Tennessee has proposed that all human cloning and all creation of embryos solely for research and destruction be banned; that the total number of embryos used for research be limited, but that embryonic stem cell research from spare embryos be approved and federally funded; and that there be increased funding for adult stem cell research. This is the sort of compromise—one that claims to be intellectually coherent, morally grounded, and practically achievable—that the anguished moderates seek.

There are many types of anguished moderates. There are morally serious pro-choicers, like representative David Wu of Oregon, who defend abortion but take concerns about the use of embryos seriously, and who realize that even the benefits of research do not justify risking a leap into a Brave New World of human cloning. There are the "soft" pro-lifers, like senator Orrin Hatch and former senator Connie Mack, who believe research on leftover frozen embryos and opposition to abortion are mutually consistent positions, since, as Hatch put it, "Life begins in the mother's womb, not in a refrigerator." Finally, there are those who believe that human cloning and research on embryonic stem cells are both wrong, but that cloning is by far the greater evil. This group is willing, if necessary, to concede some forms of embryonic stem cell research if it can draw a bright line against human cloning. It adopts, in other words, a strategy of containment, a melancholy realism about where we are and what is possible.

There will be strong pressure on both the Democratic Senate, which must decide what to do about human cloning, and President Bush, who must decide whether or not to authorize public funding for embryonic stem cell research, to come down somewhere in this anguished center.

President Bush, if one takes his earliest statements seriously, believes that research on human embryos is wrong. He assured his pro-life supporters during the campaign and in the first months of his presidency that he would not allow federal funding for research "that involves destroying living human embryos." But now he must decide whether to hold to this position, or to give in to the massive pressure to authorize at least some federal funding for embryonic stem cell research. And he must also decide how strongly to push for a ban on create-and-kill embryonic research in the private sector.

Senate majority leader Tom Daschle has a different dilemma: The Democrats risk becoming the party of human cloning. After all, Democrats in the

House voted 153 to 53 in favor of embryonic cloning; Republicans voted 194 to 25 against it. Daschle's comments after the House vote last week suggested that he is aware of this risk, and that he stands somewhere in the anguished center, if on its left-leaning, pro-research, pro-choice edge. In his statements, he went out of his way to separate the cloning debate from the stem cell debate—decrying cloning and endorsing stem cells. But what he and his party will do in the Senate is uncertain. His precise wording—"My preference is to ban cloning, period, but, you know, I also recognize that these are very, very complicated issues"—leaves some wiggle room. Will he challenge the research establishment and the plurality (perhaps even the majority) within his own party that approves of embryonic cloning? Or is health-at-any-cost the new defining principle of liberalism? Is this where the "pursuit of happiness" has taken us?

ANTI-BRAVE NEW WORLDERS

Those in the last group, which includes the authors, share a foreboding about where the new science is taking us. Its members made up the core of support for the Weldon ban on human cloning, and comprise moral conservatives (mostly religious) and some on the morally serious environmental and anti-corporate left. They imagine with horror a future that looks like Aldous Huxley's Brave New World, C. S. Lewis's abolition of man, or Pope John Paul II's culture of death. And they want to stop it.

In his brilliant critique of human cloning in the *New Republic* in May 2001, Leon Kass began with the following admonition:

> The urgency of the great political struggles of the twentieth century, successfully waged against totalitarianisms first right and then left, seems to have blinded many people to a deeper and ultimately darker truth about the present age: all contemporary societies are traveling briskly in the same utopian direction. All are wedded to the modern technological project; all march eagerly to the drums of progress and fly proudly the banner of modern science; all sing loudly the Baconian anthem, "Conquer nature, relieve man's estate."

What we are debating now is whether we have any choice in how this march turns out, whether we can stop or turn back, and whether we even want to. It is in the nature of modern democracies, certainly American democracy, that issues move in and out of sight. At present, we are in the midst of a debate on embryo research, human cloning, and stem cells. But the choices and advances that have placed these dilemmas before us did not happen overnight. They happened step by step, one innovation after the next. The dilemmas themselves were always there, if perhaps not always quite as pressing as they now seem.

Indeed, Kass's alarm in 2001 sounds similar to his warnings in the early 1970s, when he argued that the unnatural manufacture of human life in the laboratory would lead us down a path on which it would be difficult to stop. But since then, after the initial shock and horror of each new technological development, there came a period of quiet momentum in its favor, then tacit acceptance, then normalcy.

Now, the issue is publicly joined. Are there moral markers that can hold? Can we preserve the benefits of medical progress without succumbing to a post-human future? Which of our past decisions—or non-decisions—must we revisit? And how solid are the compromises of the anguished moderates? There is, in the best of these compromises, perhaps some of the prudence of those, in the first half of the nineteenth century, who thought it was enough simply to halt the spread of slavery. But as with slavery, there are inconsistencies and temptations that make the anguished moderate position unsustainable. Even if some version of Senator Frist's hair-splitting prevails, it might well turn out to be a mere Missouri Compromise, with more fundamental battles just around the corner.

For example: Any compromise built on the distinction between leftover embryos and embryos created for destruction is problematic. Couples who create scores of extra embryos at fertility clinics, and who consent for their spare embryos to be used in research, know in advance that these embryos will be used and destroyed. Certainly, this is not the couple's main purpose in creating them—any more than destruction is the main purpose of researchers who create embryos in the noble pursuit of curing disease. But in both cases, embryos are created by people who know in advance that they will be destroyed.

And what about private sector research on embryonic stem cells? If such research is morally objectionable, shouldn't it be banned, not merely deprived of federal funding? Moreover, if this work continues and succeeds, all users of modern medicine will benefit—and all will be implicated in the moral problem this "progress" raises.

Finally, even principled opponents of embryonic stem cell research and human cloning have not fully confronted the connection between the goal of relieving disease and suffering and the increasingly dehumanizing means of achieving it. Some defend doubling, tripling, quadrupling research on adult stem cells. Science itself, they say, dictates that we don't "need" embryonic stem cells, only adult ones—a point many leading scientists vehemently disagree with. And this is to say nothing of the morally problematic eugenic uses to which stem cell research—both adult and embryonic—will be put.

After all, isn't it our alleged "need" for such research that has eroded our ability to say no in the first place? Isn't it an inflamed desire for comfort, health, and longevity above all else that impels us forward, that makes us justify what initially seems unjustifiable, that blinds us to the truth about human mortality and finitude, and about the dark side of our disease-ending civilization? To cure, after all, is to eliminate, to erase, to stamp out. What begins as a quest to halt disease may end as a "compassionate" effort to stamp out the diseased themselves. And soon enough, it is not just diseases and the diseased that are a problem to be done away with, but the inconvenient and undesirable—the unintelligent, or the old, or the unfit, or those of the wrong sex.

For now, the vote in the House to say no to human cloning, to reject the modern technological project's latest Faustian bargain, is heartening. Maybe this will lead to a more fundamental democratic engagement with the threat of science and technology to human decency and human dignity. But not necessarily. Perhaps instead it will take the first live human clone to shock us fully awake. Or perhaps the emergence of the first great stem cell cure—or eugenic enhancement—will erode our resistance, and our conscience, even further, luring us all unawares toward a post-human future. But last week's vote demonstrates that such a nightmare is not inevitable.

ADDRESS TO THE NATION

President George W. Bush

Good evening. I appreciate you giving me a few minutes of your time tonight so I can discuss with you a complex and difficult issue, an issue that is one of the most profound of our time.

The issue of research involving stem cells derived from human embryos is increasingly the subject of a national debate and dinner-table discussions. The issue is confronted every day in laboratories as scientists ponder the ethical ramifications of their work. It is agonized over by parents and many couples as they try to have children, or to save children already born.

The issue is debated within the church, with people of different faiths, even many of the same faith coming to different conclusions. Many people are finding that the more they know about stem cell research, the less certain they are about the right ethical and moral conclusions.

My administration must decide whether to allow federal funds, your tax dollars, to be used for scientific research on stem cells derived from human embryos. A large number of these embryos already exist. They are the product of a process called in vitro fertilization, which helps so many couples conceive children. When doctors match sperm and egg to create life outside the womb, they usually produce more embryos than are implanted in the mother. Once a couple successfully has children, or if they are unsuccessful, the additional embryos remain frozen in laboratories.

Some will not survive during long storage; others are destroyed. A number have been donated to science and used to create privately funded stem

George W. Bush is the 43rd president of the United States. Included here is his special address to the nation on stem cell research, delivered on August 9, 2001.

cell lines. And a few have been implanted in an adoptive mother and born, and are today healthy children.

Based on preliminary work that has been privately funded, scientists believe further research using stem cells offers great promise that could help improve the lives of those who suffer from many terrible diseases—from juvenile diabetes to Alzheimer's, from Parkinson's to spinal cord injuries. And while scientists admit they are not yet certain, they believe stem cells derived from embryos have unique potential.

You should also know that stem cells can be derived from sources other than embryos—from adult cells, from umbilical cords that are discarded after babies are born, from human placentas. And many scientists feel research on these types of stem cells is also promising. Many patients suffering from a range of diseases are already being helped with treatments developed from adult stem cells.

However, most scientists, at least today, believe that research on embryonic stem cells offers the most promise because these cells have the potential to develop in all of the tissues in the body.

Scientists further believe that rapid progress in this research will come only with federal funds. Federal dollars help attract the best and brightest scientists. They ensure new discoveries are widely shared at the largest number of research facilities and that the research is directed toward the greatest public good.

The United States has a long and proud record of leading the world toward advances in science and medicine that improve human life. And the United States has a long and proud record of upholding the highest standards of ethics as we expand the limits of science and knowledge. Research on embryonic stem cells raises profound ethical questions, because extracting the stem cell destroys the embryo, and thus destroys its potential for life. Like a snowflake, each of these embryos is unique, with the unique genetic potential of an individual human being.

As I thought through this issue, I kept returning to two fundamental questions. First, are these frozen embryos human life, and therefore, something precious to be protected? And second, if they're going to be destroyed anyway, shouldn't they be used for a greater good, for research that has the potential to save and improve other lives?

I've asked those questions and others of scientists, scholars, bioethicists, religious leaders, doctors, researchers, members of Congress, my Cabinet, and my friends. I have read heartfelt letters from many Americans. I have given this issue a great deal of thought, prayer and considerable reflection. And I have found widespread disagreement.

On the first issue, are these embryos human life? Well, one researcher told me he believes this five-day-old cluster of cells is not an embryo, not yet an individual, but a pre-embryo. He argued that it has the potential for life, but it is not a life because it cannot develop in its own.

An ethicist dismissed that as a callous attempt at rationalization. Make no mistake, he told me, that cluster of cells is the same way you and I, and all the rest of us, started our lives. One goes with a heavy heart if we use these, he said, because we are dealing with the seeds of the next generation.

And to the other crucial question, if these are going to be destroyed anyway, why not use them for good purpose—I also found different answers. Many argue these embryos are byproducts of a process that helps create life, and we should allow couples to donate them to science so they can be used for good purpose instead of wasting their potential. Others will argue there's no such thing as excess life, and the fact that a living being is going to die does not justify experimenting on it or exploiting it as a natural resource.

At its core, this issue forces us to confront fundamental questions about the beginnings of life and the ends of science. It lies at a difficult moral intersection, juxtaposing the need to protect life in all its phases with the prospect of saving and improving life in all its stages.

As the discoveries of modern science create tremendous hope, they also lay vast ethical minefields. As the genius of science extends the horizons of what we can do, we increasingly confront complex questions about what we should do. We have arrived at that brave new world that seemed so distant in 1932, when Aldous Huxley wrote about human beings created in test tubes in what he called a "hatchery."

In recent weeks, we learned that scientists have created human embryos in test tubes solely to experiment on them. This is deeply troubling, and a warning sign that should prompt all of us to think through these issues very carefully.

Embryonic stem cell research is at the leading edge of a series of moral hazards. The initial stem cell researcher was at first reluctant to begin his research, fearing it might be used for human cloning. Scientists have already cloned a sheep. Researchers are telling us the next step could be to clone human beings to create individual designer stem cells, essentially to grow another you, to be available in case you need another heart or lung or liver.

I strongly oppose human cloning, as do most Americans. We recoil at the idea of growing human beings for spare body parts, or creating life for our convenience. And while we must devote enormous energy to conquering disease, it is equally important that we pay attention to the moral concerns raised by the new frontier of human embryo stem cell research. Even the most noble ends do not justify any means.

My position on these issues is shaped by deeply held beliefs. I'm a strong supporter of science and technology, and believe they have the potential for incredible good—to improve lives, to save life, to conquer disease. Research offers hope that millions of our loved ones may be cured of a disease and rid of their suffering. I have friends whose children suffer from juvenile diabetes. Nancy Reagan has written me about President Reagan's struggle with Alzheimer's. My own family has confronted the tragedy of childhood leukemia. And, like all Americans, I have great hope for cures.

I also believe human life is a sacred gift from our Creator. I worry about a culture that devalues life, and believe as your President I have an important obligation to foster and encourage respect for life in America and throughout the world. And while we're all hopeful about the potential of this research, no one can be certain that the science will live up to the hope it has generated.

Eight years ago, scientists believed fetal tissue research offered great hope for cures and treatments—yet, the progress to date has not lived up to its initial expectations. Embryonic stem cell research offers both great promise and great peril. So I have decided we must proceed with great care.

As a result of private research, more than 60 genetically diverse stem cell lines already exist. They were created from embryos that have already been destroyed, and they have the ability to regenerate themselves indefinitely, creating ongoing opportunities for research. I have concluded that we should allow federal funds to be used for research on these existing stem cell lines, where the life-and-death decision has already been made.

Leading scientists tell me research on these 60 lines has great promise that could lead to breakthrough therapies and cures. This allows us to explore the promise and potential of stem cell research without crossing a fundamental moral line, by providing taxpayer funding that would sanction or encourage further destruction of human embryos that have at least the potential for life.

I also believe that great scientific progress can be made through aggressive federal funding of research on umbilical cord, placenta, adult and animal stem cells which do not involve the same moral dilemma. This year, your government will spend $250 million on this important research.

I will also name a President's council to monitor stem cell research, to recommend appropriate guidelines and regulations, and to consider all of the medical and ethical ramifications of biomedical innovation. This council will consist of leading scientists, doctors, ethicists, lawyers, theologians and others, and will be chaired by Dr. Leon Kass, a leading biomedical ethicist from the University of Chicago. This council will keep us apprised of new developments and give our nation a forum to continue to discuss and evaluate these important issues.

As we go forward, I hope we will always be guided by both intellect and heart, by both our capabilities and our conscience.

I have made this decision with great care, and I pray it is the right one.

Thank you for listening. Good night, and God bless America.

STEM CELL SCIENCE AND THE PRESERVATION OF LIFE

President George W. Bush

Some of the hardest ethical decisions pit good against good. In the case of stem cells, the promise of miracle cures is set against the protection of developing life. The conflict has left Americans divided, even in their own minds.

Stem cell research is still at an early, uncertain stage, but the hope it offers is amazing: infinitely adaptable human cells to replace damaged or defective tissue and treat a wide variety of diseases.

Yet the ethics of medicine are not infinitely adaptable. There is at least one bright line: We do not end some lives for the medical benefit of others. For me, this is a matter of conviction: a belief that life, including early life, is biologically human, genetically distinct and valuable. But one need not be pro-life to be disturbed by the prospect of fetal farming or cloning to provide spare human parts. Most Americans share a belief that human life should not be reduced to a tool or a means.

There are, however, two ways for the federal government to aggressively promote stem cell research without inviting ethical abuses.

First, we can encourage research on stem cells removed from sources other than embryos: adult cells, umbilical cords and human placentas. Many researchers see great potential in these cells—and they have already been used to develop several new therapies.

Second, we can encourage research on embryonic stem cell lines that already exist. These cells can reproduce themselves in the laboratory, perhaps indefinitely. Stem cell lines at the University of Wisconsin have been

This essay was originally published in the *New York Times* on August 12, 2001.

producing cells for over two years. More than 60 of these cell lines now exist around the world. According to the National Institutes of Health, these lines are genetically diverse and sufficient in number for the research ahead.

Therefore my administration has adopted the following policy: Federal funding for research on existing stem cell lines will move forward; federal funding that sanctions or encourages the destruction of additional embryos will not. While it is unethical to end life in medical research, it is ethical to benefit from research where life and death decisions have already been made.

There is a precedent. The only licensed live chickenpox vaccine used in the United States was developed, in part, from cells derived from research involving human embryos. Researchers first grew the virus in embryonic lung cells, which were later cloned and grown in two previously existing cell lines. Many ethical and religious leaders agree that even if the history of this vaccine raises ethical questions, its current use does not.

Stem cell research takes place on a slippery slope of moral concern where much biomedical research is and will be conducted. We must keep our ethical footing. Government has a clear duty to promote scientific discovery—and a duty to define certain boundaries:

Under my policy, existing stem cell lines, to be used in publicly supported research, must be derived (1) with the informed consent of donors, (2) from excess embryos created solely for reproductive purposes and (3) without any financial inducements to the donors.

I have directed the National Institutes of Health to establish a national human embryonic stem cell registry. This will ensure that ethical research standards are observed by all recipients of federal funding.

Soon I will appoint a Presidential Council on Bioethics, chaired by Dr. Leon Kass, to advise my administration on moral and scientific questions raised by biomedical research. My administration supports legislative efforts to prohibit the cloning of human beings for any purpose, and also to prohibit the production of human embryos solely to be destroyed in medical research.

As we enter the new territory of modern science, the choices will only grow more difficult. The new technologies we create—with their potential to cure disease and relieve suffering—may well define our age. But we will also be defined by the care and sense of self-restraint and responsibility with which we took up these new powers.

Power—even technological power—is always judged by its ends and its means. Seeking noble ends by any means is unacceptable when life itself is in the balance.

Biomedical progress should be welcomed, promoted and funded—yet it can and must be humanized. Caution is demanded, because second thoughts will come too late. As we work to extend our lives, we must do so in ways that preserve our humanity.

STEM CELLS

BUSH'S BROKEN PROMISE

Kenneth L. Connor

On Thursday night President Bush announced his long-awaited decision on the use of taxpayer funds for research on human embryos. The impact of his decision on the fate of hundreds of thousands of embryos and on progress toward medical treatments will not be clear for months or even decades. Its impact on the character of his presidency, however, is clear now: He has made a breach of faith in the service of an untenable compromise.

The promise that President Bush broke in his nationally televised address, the first by a U.S. president devoted to any sanctity-of-life issue, was no mere campaign throwaway. As recently as May 18, in a letter to the Culture of Life Foundation, the president made his views plain: "I oppose federal funding for stem cell research that involves destroying living human embryos. I support innovative medical research on life-threatening and debilitating diseases, including promising research on stem cells from adult tissue."

The president's announcement attempts to finesse his May 18 statement by distinguishing "involves" from "involved." Taxpayer funds will be used for research only on cell lines derived from embryos already killed. No cell lines derived from fresh killing—presumably, that is, killing after the date of the president's speech—will be permitted.

Embryo research advocates are already challenging the president's numbers. They say that there are far fewer stem cell lines in existence than his estimate of 60. They know that once the principle is given away, we are only haggling over the price. If 30 more stem cell lines are created in the private

Kenneth L. Connor is president of the Family Research Council and a leader in the pro-life movement. This essay originally appeared in the *Washington Post* on August 11, 2001.

sector by killing human embryos in the next six months, the federal government will not have been "involved" in their demise either. On what principle will the president refuse to authorize use of these fresher human cells?

The issue isn't academic. Existing stem cell lines are believed to have a "shelf life" of some two years. Beyond that, after multiple cell divisions in man-made environments, the cells deteriorate. No serious person—check the capital markets for capital flows to biotech firms—believes that the therapeutic potential of stem cells will be thoroughly explored two years from now. The president's new "bright line" for federal involvement will collapse.

At the end of the day, however, this issue has never been about the quantity or sequence of embryo destruction and experimentation. It has been about this principle: Can a human being cease to have value for himself or herself and merely become a means to preserve life and health for others? The stage of development of these human beings renders them especially vulnerable, but it does not alter their status or the grave importance of the president's decision.

Why indeed have we paid so much attention to the fate of such tiny human beings? Because even though they are small, the ethical principle at stake is huge. For 3,000 years, the first rule of medicine has been "Do No Harm." By abandoning that rule, the president has helped to usher in a new era marked by the philosophy that the ends justify the means. As Bush even acknowledged, Aldous Huxley's vision of a "Brave New World" has arrived. We now say that it is permissible to kill so long as we intend to bring good from it. The new *modus operandi* for medicine will be "kill to cure." This was the ethos of Dr. Mengele, who experimented on doomed twins at Auschwitz.

Opposition to experiments on embryonic children has been informed by the deep belief that people are not products to be used for the benefit of others. No commercial gain or scientific benefit can justify the slaughter of the innocent. "Nothing stamped in the divine image," said Lincoln, "was sent into the world to be trod upon." America suffered a terrible Civil War, with the loss of 620,000 lives, to reestablish that principle.

For seven months now we have called upon the president to honor his pledge—the commitments made by his party's platform and his own campaign—to protect innocent human life. His change of course will forever be a blot upon his record in office.

BUSH'S STEM-CELL RULING

A MISSOURI COMPROMISE

Eric Cohen

Only a sage could have predicted that President Bush's first major decision would be about federal funding of embryonic stem-cell research. For the first few months of his presidency, most people expected stem cells to be a one-day story—like federal funding of overseas abortion clinics—with the usual back-and-forth between the cultural right and left. But this was not to be.

The issue of stem cells struck a national nerve, bringing into sharp relief deep divides about right and wrong, life and death, and the meaning and source of human dignity. It introduced, in embryonic form, what may turn out to be the new bloody crossroads of American politics: where giant leaps forward in medical science meet deeply entrenched differences about what makes life sacred, and where the American gospel of progress meets the biblical admonition against human pride.

President Bush's address to the nation last Thursday, with great honesty and sobriety, laid out these differences and the competing goods (and evils) that underlie them. He described the "unique potential" of embryonic stem cells "to save life," but also the prospect of a "brave new world" of "spare body parts" and "designer stem cells" made through human cloning. With the humility that he has tried to make the moral mark of his presidency, he described what is likely an accurate portrait of the American middle: "Many people are finding that the more they know about stem-cell research, the less certain they are about the right ethical and moral conclusions."

This essay was originally published in the *Los Angeles Times* on August 12, 2001.

Then he offered a compromise: Federal funding for research on embryonic stem-cell lines that have already been created, where, as he put it, "the life-and-death decision has already been made."

This compromise was built on three principles: One, that human embryos are life—not to be discarded, harvested, cloned, or used simply "for our convenience"; two, that the hope of the sick and suffering for cures should be kindled, that the march of medical progress must continue, but that "even the most noble ends do not justify any means"; and that the government should not take life for research, but it should fund research where others have already taken life—since there is no turning back the clock, and since the memory of destroyed embryos may best be honored by giving hope to those who might still live in the here and now.

It was a statesmanlike speech and a statesmanlike proposal, a compromise that aimed to be at once morally serious, politically viable and practically sustainable. But whether the Bush compromise can hold is far from certain, for many reasons.

First, the destruction of embryos for research will continue apace in the private sector. Which raises the question: If research that involves the creation of embryos solely for research and destruction is immoral, shouldn't there be a ban on all such research, not just on the federal funding of it? And if the guiding principle of the Bush compromise is to fund research on embryos that have already been destroyed, should the government fund research on future stem-cell lines, since, inevitably, private researchers will continue to make such life-and-death decisions in the months and years ahead?

Second, if this research is successful, won't it become a regular part of modern medicine, making it hard for those who believe such research is wrong to live according to their values while remaining fully integrated in American life? Is it possible that just as a separate society of home-schoolers has grown, a separate society of "home-healers" might grow, and with it the already deep moral rift in the nation?

Third, what happens if the 60 or so existing stem-cell lines are "not enough," as many leading scientists are already claiming? Will President Bush revisit his decision? Are the moral boundaries he has drawn strong enough—and widely accepted enough by the nation at large—to hold up against the next wave of medical promises?

For now, Bush's decision seems to have achieved what most believed to be impossible: approval from many members of both the pro-life community and the patient-advocacy and medical research community. But this may turn out to be a Missouri Compromise—an effort to find the best possible solution, for now, with larger debates and disagreements just around the corner.

Throughout the campaign and during his first six months in office, the president vowed to bring a new civility and decency to American public life. After his legislative victories on tax cuts, energy and the patient's bill of rights, his staff said he would turn his attention to America's values, including initiatives like urging the media to put more good news in the newspaper and urging kids to e-mail their grandparents. Small and silly as these initiatives seem, they are part of a larger Bush project: to promote values without inciting conflict, to make America more virtuous without opening up the Pandora's box of profound disagreement about what it is exactly that makes America virtuous. On a much more serious and sobering level, the president tried to accomplish this feat with stem cells: to preserve America's shared values while reining in the nation's deepest moral divides. It is, at best, an honorable enterprise, but whether it will prove lasting or significant, especially in the face of the looming challenges of the genetic revolution, is a great unknown.

After the speech, White House Chief of Staff Andrew Card described Bush's decision as "perfect for America." Perhaps more accurately, it was a decision that perfectly reflects America's uncertainty. A CNN/USA Today poll showed that 54 percent of Americans believe embryonic stem-cell research is morally wrong, but 69 percent believe it is medically necessary. In other words, most Americans believe it is necessary to do wrong, that it is morally right (or at least morally justifiable) to do evil. This may be the subtlety required by such life-and-death questions—the same subtlety required when good nations decide to go to war. Or it may, in the end, be an untenable—and unsustainable—moral confusion.

PART III

Mortality and the American Character

WHY NOT IMMORTALITY?

Leon R. Kass

If life is good and more is better, should we not regard death as a disease and try to cure it? Although this formulation of the question may seem too futuristic or far-fetched, there are several reasons for taking it up and treating it seriously.

First, reputable scientists are today answering the question in the affirmative and are already making large efforts toward bringing about a cure. Three kinds of research, still in their infancy, are attracting new attention and energies. First is the use of hormones, especially human growth hormone (hGH), to restore and enhance youthful bodily vigor. In the United States, over ten thousand people—including many physicians—are already injecting themselves daily with hGH for anti-aging purposes, with apparently remarkable improvements in bodily fitness and performance, though there is as yet no evidence that the hormones yield any increase in life expectancy. When the patent on hGH expires in 2002 and the cost comes down from its current $1,000 per month, many more people are almost certainly going to be injecting themselves from the hormonal fountain of youth.

Second is research on stem cells, those omnicompetent primordial cells that, on different signals, turn into all the different differentiated tissues of the body—liver, heart, kidney, brain, etc. Stem cell technologies—combined with techniques of cloning—hold out the promise of an indefinite supply of replacement tissues and organs for any and all worn-out body parts. This is a booming area in commercial biotechnology, and one of the leading biotech entrepreneurs has been touting his company's research as promising indefinite prolongation of life.

Selections from "L'Chaim and Its Limits: Why Not Immortality?" *First Things,* May 2000.

Third, there is research into the genetic switches that control the biological processes of aging. The maximum life span for each species—roughly one hundred years for human beings—is almost certainly under genetic control. In a startling recent discovery, fruit-fly geneticists have shown that mutations in a *single* gene produce a 50 percent increase in the natural lifetime of the flies. Once the genes involved in regulating the human life cycle and setting the midnight hour are identified, scientists predict that they will be able to increase the human maximum age well beyond its natural limit. Quite frankly, I find some of the claims and predictions to be overblown, but it would be foolhardy to bet against scientific and technical progress along these lines.

But even if cures for aging and death are a long way off, there is a second and more fundamental reason for inquiring into the radical question of the desirability of gaining a cure for death. For truth to tell, victory over mortality is the unstated but implicit goal of modern medical science, indeed of the entire modern scientific project, to which mankind was summoned almost four hundred years ago by Francis Bacon and René Descartes. They quite consciously trumpeted the conquest of nature for the relief of man's estate, and they founded a science whose explicit purpose was to reverse the curse laid on Adam and Eve, and especially to restore the tree of life, by means of the tree of (scientific) knowledge. With medicine's increasing successes, realized mainly in the last half century, every death is increasingly regarded as premature, a failure of today's medicine that future research will prevent. In parallel with medical progress, a new moral sensibility has developed that serves precisely medicine's crusade against mortality: anything is permitted if it saves life, cures disease, prevents death. Regardless, therefore, of the imminence of anti-aging remedies, it is most worthwhile to reexamine the assumption upon which we have been operating: that everything should be done to preserve health and prolong life as much as possible, and that all other values must bow before the biomedical gods of better health, greater vigor, and longer life.

Recent proposals that we should conquer aging and death have not been without their critics. The criticism takes two forms: predictions of bad social consequences and complaints about distributive justice. Regarding the former, there are concerns about the effect on the size and age distribution of the population. How will growing numbers and percentages of people living well past one hundred affect, for example, work opportunities, retirement plans, hiring and promotion, cultural attitudes and beliefs, the structure of family life, relations between the generations, or the locus of rule and authority in government, business, and the professions? Even the most cursory examination of these matters suggests that the cumulative results of aggregated decisions for longer and more vigorous life could be highly disruptive and undesirable, even to the point that many individuals would be *worse off* through most of their lives, and worse off enough to offset the benefits of better health afforded them near the end of life. Indeed, several people have predicted that retardation of aging will present a classic instance of the Tragedy of the Commons, in which genuine and sought-for gains to individuals are nullified or worse, owing to the social consequences of granting them to everyone.

But other critics worry that technology's gift of long or immortal life will not be granted to everyone, especially if, as is likely, the treatments turn out to be expensive. Would it not be the ultimate injustice if only some people could afford a deathless existence, if the world were divided not only into rich and poor but into mortal and immortal?

Against these critics, the proponents of immortality research answer confidently that we will gradually figure out a way to solve these problems. We can handle any adverse social consequences through careful planning; we can overcome the inequities through cheaper technologies. Though I think these optimists woefully naive, let me for the moment grant their view regarding these issues. For both the proponents and their critics have yet to address thoughtfully the heart of the matter, the question of the goodness of the goal. The core question is this: Is it really true that longer life for individuals is an unqualified good?

How *much* longer life is a blessing for an individual? Ignoring now the possible harms flowing back to individuals from adverse social consequences, how much more life is good for us as individuals, other things being equal? How much more life do we want, assuming it to be healthy and vigorous? Assuming that it were up to us to set the human life span, where would or should we set the limit and why?

The simple answer is that no limit should be set. Life is good, and death is bad. Therefore, the more life the better, provided, of course, that we remain fit and our friends do, too.

This answer has the virtues of clarity and honesty. But most public advocates of conquering aging deny any such greediness. They hope not for immortality, but for something reasonable—just a few more years.

How many years are reasonably few? Let us start with ten. Which of us would find unreasonable or unwelcome the addition of ten healthy and vigorous years to his or her life, years like those between ages thirty and forty?

We could learn more, earn more, see more, do more. Maybe we should ask for five years on top of that? Or ten? Why not fifteen, or twenty, or more?

If we can't immediately land on the reasonable number of added years, perhaps we can locate the principle. What is the principle of reasonableness? Time needed for our plans and projects yet to be completed? Some multiple of the age of a generation, say, that we might live to see great-grandchildren fully grown? Some notion—traditional, natural, revealed—of the proper life span for a being such as man? We have no answer to this question. We do not even know how to choose among the principles for setting our new life span.

Under such circumstances, lacking a standard of reasonableness, we fall back on our wants and desires. Under liberal democracy, this means the desires of the majority for whom the attachment to life—or the fear of death—knows no limits. It turns out that the simple answer is the best: we want to live and live, and not to wither and not to die. For most of us, especially under modern secular conditions in which more and more people believe that this is the only life they have, the desire to prolong the life span (even modestly) must be seen as expressing a desire *never* to grow old and die. However naive their counsel, those who propose immortality deserve credit: they honestly and shamelessly expose this desire.

Some, of course, eschew any desire for longer life. They seek not adding years to life, but life to years. For them, the ideal life span would be our natural (once thought three-, now known to be) fourscore and ten, or if by reason of strength, fivescore, lived with full powers right up to death, which could come rather suddenly, painlessly, at the maximal age.

This has much to recommend it. Who would not want to avoid senility, crippling arthritis, the need for hearing aids and dentures, and the degrading dependencies of old age? But, in the absence of these degenerations, would we remain content to spurn longer life? Would we not become even more disinclined to exit? Would not death become even more of an affront? Would not the fear and loathing of death increase in the absence of its harbingers? We could no longer comfort the widow by pointing out that her husband was delivered from his suffering. Death would always be untimely, unprepared for, shocking.

Montaigne saw it clearly:

> I notice that in proportion as I sink into sickness, I naturally enter into a certain disdain for life. I find that I have much more trouble digesting this resolution when I am in health than when I have a fever. Inasmuch as I no longer cling so hard to the good things of life when I begin to lose the use and plea-

> sure of them, I come to view death with much less frightened eyes. This makes me hope that the farther I get from life and the nearer to death, the more easily I shall accept the exchange. . . . If we fell into such a change [decrepitude] suddenly, I don't think we could endure it. But when we are led by Nature's hand down a gentle and virtually imperceptible slope, bit by bit, one step at a time, she rolls us into this wretched state and makes us familiar with it; so that we find no shock when youth dies within us, which in essence and in truth is a harder death than the complete death of a languishing life or the death of old age; inasmuch as the leap is not so cruel from a painful life as from a sweet and flourishing life to a grievous and painful one.

Thus it is highly likely that even a modest prolongation of life with vigor or even only a preservation of youthfulness with no increase in longevity would make death less acceptable and would exacerbate the desire to keep pushing it away—unless, for some reason, such life could also prove less satisfying.

Could longer, healthier life be less satisfying? How could it be, if life is good and death is bad? Perhaps the simple view is in error. Perhaps mortality is not simply an evil, perhaps it is even a blessing—not only for the welfare of the community, but even for us as individuals. How could this be?

I wish to make the case for the virtues of mortality. Against my own strong love of life, and against my even stronger wish that no more of my loved ones should die, I aspire to speak truth to my desires by showing that the finitude of human life is a blessing for every human individual, whether he knows it or not.

I know I won't persuade many people to my position. But I do hope I can convince readers of the gravity—I would say, the unique gravity—of this question. We are not talking about some minor new innovation with ethical wrinkles about which we may chatter or regulate as usual. Conquering death is not something that we can try for a while and then decide whether the results are better or worse—according to, God only knows, what standard. On the contrary, this is a question in which our very humanity is at stake, not only in the consequences but also in the very meaning of the choice. For to argue that human life would be better without death is, I submit, to argue that human life would be better being something other than human. To be immortal would not be just to continue life as we mortals now know it, only forever. The new immortals, in the decisive sense, would not be like us at all. If this is true, a human choice for bodily immortality would suffer from the deep confusion of choosing to have some great good only on the condition of turning into someone else. Moreover, such an immortal someone else, in my view, will be less well off than we mortals are now, thanks indeed to our mortality.

It goes without saying that there is no virtue in the death of a child or a young adult, or the untimely or premature death of anyone, before they had attained to the measure of man's days. I do not mean to imply that there is virtue in the particular *event* of death for anyone. Nor am I suggesting that separation through death is not painful for the survivors, those for whom the deceased was an integral part of their lives. Instead, my question concerns the fact of our finitude, the fact of our mortality—the fact *that we must die,* the fact that a full life for a human being has a biological, built-in limit, one that has evolved as part of our nature. Does this fact also have value? Is our finitude good for us—as individuals? (I intend this question entirely in the realm of natural reason and apart from any question about a life after death.)

To praise mortality must seem to be madness. If mortality is a blessing, it surely is not widely regarded as such. Life seeks to live, and rightly suspects all counsels of finitude. "Better to be a slave on earth than the king over all the dead," says Achilles in Hades to the visiting Odysseus, in apparent regret for his prior choice of the short but glorious life. Moreover, though some cultures—such as the Eskimo—can instruct and moderate somewhat the lust for life, liberal Western society gives it free rein, beginning with a political philosophy founded on a fear of violent death, and reaching to our current cults of youth and novelty, the cosmetic replastering of the wrinkles of age, and the widespread anxiety about disease and survival. Finally, the virtues of finitude—if there are any—may never be widely appreciated in any age or culture, if appreciation depends on a certain wisdom, if wisdom requires a certain detachment from the love of oneself and one's own, and if the possibility of such detachment is given only to the few. Still, if it is wisdom, the rest of us should hearken, for we may learn something of value for ourselves.

How, then, might our finitude be good for us? I offer four benefits, first among which is *interest and engagement*. If the human life span were increased even by only twenty years, would the pleasures of life increase proportionately? Would professional tennis players really enjoy playing 25 percent more games of tennis? Would the Don Juans of our world feel better for having seduced 1,250 women rather than 1,000? Having experienced the joys and tribulations of raising a family until the last had left for college, how many parents would like to extend the experience by another ten years? Likewise, those whose satisfaction comes from climbing the career ladder might well ask what there would be to do for fifteen years after one had been CEO of Microsoft, a member of Congress, or the President of Harvard for a quarter of a century? Even less clear are the additions to personal happiness from more of the same of the less pleasant and less fulfilling activities in which so many of us are engaged so much of the time. It seems to be as the poet says: "We move and ever spend our lives amid the same things, and not by any length of life is any new pleasure hammered out."

Second, *seriousness and aspiration*. Could life be serious or meaningful without the limit of mortality? Is not the limit on our time the ground of our taking life seriously and living it passionately? To know and to feel that one goes around only once, and that the deadline is not out of sight, is for many people the necessary spur to the pursuit of something worthwhile. "Teach us to number our days," says the Psalmist, "that we may get a heart of wisdom." To number our days is the condition for making them count. Homer's immortals—Zeus and Hera, Apollo and Athena—for all their eternal beauty and youthfulness, live shallow and rather frivolous lives, their passions only transiently engaged, in first this and then that. They live as spectators of the mortals, who by comparison have depth, aspiration, genuine feeling, and hence a real center in their lives. Mortality makes life matter.

There may be some activities, especially in some human beings, that do not require finitude as a spur. A powerful desire for understanding can do without external proddings, let alone one related to mortality; and as there is never too much time to learn and to understand, longer, more vigorous life might be simply a boon. The best sorts of friendship, too, seem capable of indefinite growth, especially where growth is somehow tied to learning—though one may wonder whether real friendship doesn't depend in part on the shared perceptions of a common fate. But, in any case, I suspect that these are among the rare exceptions. For most activities, and for most of us, I think it is crucial that we recognize and feel the force of not having world enough and time.

A third matter, *beauty and love*. Death, says Wallace Stevens, is the mother of beauty. What he means is not easy to say. Perhaps he means that only a mortal being, aware of his mortality and the transience and vulnerability of all natural things, is moved to make beautiful artifacts, objects that will last, objects whose order will be immune to decay as their maker is not, beautiful objects that will bespeak and beautify a world that needs beautification, beautiful objects for other mortal beings who can appreciate what they cannot themselves make because of a taste for the beautiful, a taste perhaps connected to awareness of the ugliness of decay.

Perhaps the poet means to speak of natural beauty as well, which beauty—unlike that of objects of art—depends on its *im*permanence. Could the beauty of flowers depend on the fact that they will soon wither? Does the beauty of spring warblers depend upon the fall drabness that precedes and follows? What about the fading, late afternoon winter light or the spreading sunset? Is the beautiful necessarily fleeting, a peak that cannot be sustained? Or does the poet mean not that the beautiful is beautiful because mortal, but that our appreciation of its beauty depends on our appreciation of mortality—in us and in the beautiful? Does not love swell before the beautiful precisely on recognizing that it (and we) will not always be? Is not our mortality the cause of our enhanced appreciation of the beautiful and the worthy and of our treasuring and loving them? How deeply could one deathless "human" being love another?

Fourth, there is the peculiarly human beauty of character, *virtue and moral excellence*. To be mortal means that it is possible to give one's life, not only in one moment, say, on the field of battle, but also in the many other ways in which we are able in action to rise above attachment to survival. Through moral courage, endurance, greatness of soul, generosity, devotion to justice—in acts great and small—we rise above our mere creatureliness, spending the precious coinage of the time of our lives for the sake of the noble and the good and the holy. We free ourselves from fear, from bodily pleasures, or from attachments to wealth—all largely connected with survival—and in doing virtuous deeds overcome the weight of our neediness; yet for this nobility, vulnerability and mortality are the necessary conditions. The immortals cannot be noble.

Of this, too, the poets teach. Odysseus, long suffering, has already heard the shade of Achilles' testimony in praise of life when he is offered immortal life by the nymph Calypso. She is a beautiful goddess, attractive, kind, yielding; she sings sweetly and weaves on a golden loom; her island is well-ordered and lovely, free of hardships and suffering. Says the poet, "Even a god who came into that place would have admired what he saw, the heart delighted within him." Yet Odysseus turns down the offer to be lord of her household and immortal:

> Goddess and queen, do not be angry with me. I myself know that all you say is true and that circumspect Penelope can never match the impression you make for beauty and stature. She is mortal after all, and you are immortal and ageless. But even so, what I want and all my days I pine for is to go back to my house and see that day of my homecoming. And if some god batters me far out on the wine-blue water, I will endure it, keeping a stubborn spirit inside me, for already I have suffered much and done much hard work on the waves and in the fighting.

To suffer, to endure, to trouble oneself for the sake of home, family, community, and genuine friendship, is truly to live, and is the clear choice of this exemplary mortal. This choice is both the mark of his excellence and the basis for the visible display of his excellence in deeds noble and just. Immortality is a kind of oblivion—like death itself.

But, someone might reasonably object, if mortality is such a blessing, why do so few cultures recognize it as such? Why do so many teach the promise of life after death, of something eternal, of something imperishable? This takes us to the heart of the matter.

What is the meaning of this concern with immortality? *Why* do we human beings seek immortality? Why do we want to live longer or forever? Is it really first and most because we do not want to die, because we do not want to leave this embodied life on earth or give up our earthly pastimes, because we want to see more and do more? I do not think so. This may be what we say, but it is not what we finally mean. Mortality as such is not our defect, nor bodily immortality our goal. Rather, mortality is at most a pointer, a derivative manifestation, or an accompaniment of some deeper deficiency. The promise of immortality and eternity answers rather to a deep truth about the human soul: the human soul yearns for, longs for, aspires to some condition, some state, some goal toward which our earthly activities are directed but which cannot be attained in earthly life. Our soul's reach exceeds our grasp; it seeks more than continuance; it reaches for something beyond us, something that for the most part eludes us. Our distress with mortality is the derivative manifestation of the conflict between the transcendent longings of the soul and the all-too-finite powers and fleshly concerns of the body.

What is it that we lack and long for, but cannot reach? One possibility is completion in another person. For example, Plato's Aristophanes says we seek wholeness through complete and permanent bodily and psychic union with a unique human being whom we love, our "missing other half." Plato's Socrates, in contrast, says it is rather wholeness through wisdom, through comprehensive knowledge of the beautiful truth about the whole, that which philosophy seeks but can never attain. Yet again, biblical religion says we seek wholeness through dwelling in God's presence, love, and redemption—a restoration of innocent wholeheartedness lost in the Garden of Eden. But, please note, these and many other such accounts of human aspiration, despite their differences, all agree on this crucial point: man longs not so much for deathlessness as for wholeness, wisdom, goodness, and godliness—longings that cannot be satisfied fully in our embodied earthly life, the only life, by natural reason, we know we have. Hence the attractiveness of any prospect or promise of a different and thereby fulfilling life hereafter. The decisive inference is clear: none of these longings can be answered by prolonging earthly life. Not even an unlimited amount of "more of the same" will satisfy our deepest aspirations.

If this is correct, there follows a decisive corollary regarding the battle against death. The human taste for immortality, for the imperishable and the eternal, is not a taste that the biomedical conquest of death could satisfy. We would still be incomplete; we would still lack wisdom; we would still lack God's presence and redemption. Mere continuance will not buy fulfillment. Worse, its pursuit threatens—already threatens—human happiness by distracting us from the goals toward which our souls naturally point. By diverting our aim, by misdirecting so much individual and social energy toward the goal of bodily immortality, we may seriously undermine our chances for living as well as we can and for satisfying to some extent, however incompletely, our deepest longings for what is best. The implication for human life is hardly nihilistic: once we acknowledge and accept our finitude, we can concern ourselves with living *well*, and care first and most for the *well-being* of our souls, and not so much for their mere existence.

But perhaps this is all a mistake. Perhaps there is no such longing of the soul. Perhaps there is no soul. Certainly modern science doesn't speak about the soul; neither does medicine or even our *psych*iatrists, whose name means "healers of the soul." Perhaps we are just animals, complex ones to be sure, but animals nonetheless, content just to be here, frightened in the face of danger, avoiding pain, seeking pleasure.

Curiously, however, biology has its own view of our nature and its inclinations. Biology also teaches about transcendence, though it eschews talk about the soul. Biology has long shown us a feasible way to rise above our finitude and to participate in something permanent and eternal: I refer not to stem cells, but to procreation—the bearing and caring for offspring, for the sake of which many animals risk and even sacrifice their lives. Indeed, in all higher animals, reproduction *as such* implies both the acceptance of the death of self and participation in its transcendence. The salmon, willingly swimming upstream to spawn and die, makes vivid this universal truth.

But man is natured for more than spawning. Human biology teaches how our life points beyond itself—to our offspring, to our community, to our species. Like the other animals, man is built for reproduction. More than the other animals, man is also built for sociality. And, alone among the animals, man is also built for culture—not only though capacities to transmit and receive skills and techniques, but also through capacities for shared beliefs, opinions, rituals, traditions. We are built with leanings toward, and capacities for, perpetuation. Is it not possible that aging and mortality are part of this construction, and that the rate of aging and the human life span have been selected for their usefulness to the task of perpetuation? Could not extending the human life span place a great strain on our nature, jeopardizing our project and depriving us of success? Interestingly, perpetuation is a goal that *is* attainable, a transcendence of self that *is* (largely) realizable. Here is a form of participating in the enduring that is open to us, without qualification—provided, that is, that we remain open to it.

Biological considerations aside, simply to covet a prolonged life span for ourselves is both a sign and a cause of our failure to open ourselves to procreation and to any higher purpose. It is probably no accident that it is a generation whose intelligentsia proclaim the death of God and the meaninglessness of life that embarks on life's indefinite prolongation and that seeks to cure the emptiness of life by extending it forever. For the desire to prolong youthfulness is not only a childish desire to eat one's life and keep it; it is also an expression of a childish and narcissistic wish incompatible with devotion to posterity. It seeks an endless present, isolated from anything truly eternal, and severed from any true continuity with past and future. It is in principle hostile to children, because children, those who come after, are those who will take one's place; *they* are life's answer to mortality, and their presence in one's house is a constant reminder that one no longer belongs to the frontier generation. One cannot pursue agelessness for oneself and remain faithful to the spirit and meaning of perpetuation.

In perpetuation, we send forth not just the seed of our bodies, but also the bearer of our hopes, our truths, and those of our tradition. If our children are to flower, we need to sow them well and nurture them, cultivate them in rich and wholesome soil, clothe them in fine and decent opinions and mores, and direct them toward the highest light, to grow straight and tall—that they may take our place as we took that of those who planted us and made way for us, so that in time they, too, may make way and plant. But if they are truly to flower, we must go to seed; we must wither and give ground.

Against these considerations, the clever ones will propose that if we could do away with death, we would do away with the need for posterity. But that is a self-serving and shallow answer, one that thinks of life and aging solely in terms of the state of the body. It ignores the psychological effects simply of the passage of time—of experiencing and learning about the way things are. After a while, no matter how healthy we are, no matter how respected and well placed we are socially, most of us cease to look upon the world with fresh eyes. Little surprises us, nothing shocks us, righteous indignation at injustice dies out. We have seen it all already, seen it all. We have often been deceived, we have made many mistakes of our own. Many of us become small-souled, having been humbled not by bodily decline or the loss of loved ones but by life itself. So our ambition also begins to flag, or at least our noblest ambitions. As we grow older, Aristotle already noted, we "aspire to nothing great and exalted and crave the mere necessities and comforts of existence." At some point, most of us turn and say to our intimates, Is this all there is? We settle, we accept our situation—if we are lucky enough to be able to accept it. In many ways, perhaps in the most profound ways, most of us go to sleep long before our deaths—and we might even do so earlier in life if death no longer spurred us to make something of ourselves.

In contrast, it is in the young where aspiration, hope, freshness, boldness, and openness spring anew—even when they take the form of overturning our monuments. Immortality for oneself through children may be a delusion, but participating in the natural and eternal renewal of human possibility through children is not—not even in today's world.

For it still stands as it did when Homer made Glaukos say to Diomedes:

> As is the generation of leaves, so is that of humanity. The wind scatters the leaves to the ground, but the live timber burgeons with leaves again in the season of spring returning. So one generation of man will grow while another dies.

And yet it also still stands, as this very insight of Homer's itself reveals, that human beings are in another respect unlike the leaves; that the eternal renewal of human beings embraces also the eternally human possibility of learning and self-awareness; that we, too, here and now may participate with Homer, with Plato, with the Bible, yes with Descartes and Bacon, in catching at least some glimpse of the enduring truths about nature, God, and human affairs; and that we, too, may hand down and perpetuate this pursuit of wisdom and goodness to our children and our children's children. Children and their education, not growth hormone and perpetual organ replacement, are life's—and wisdom's—answer to mortality.

BIODEMOCRACY IN AMERICA

Adam Wolfson

Recent breakthroughs in biotechnology have led to all kinds of speculation about America's future in the twenty-first century, the "biotech century." The discussion has been lively and entertaining, even informative at times, but the fact is that predicting the future is a fool's game. The collapse of communism in 1989 caught us as unaware as the terrorist assault on America in 2001. Nor does our track record improve when the relevant variables are well known and quantifiable. A few months before the 2000 election, several prominent political scientists, using state-of-the-art modeling techniques, predicted that Al Gore would defeat George W. Bush by a landslide. If we lack the ability to forecast events only a few months off, surely it would be ludicrous to attempt to divine how as-yet-unknown discoveries in such complex fields as biochemistry and genomics will affect our society.

That biotechnology's advance will affect our politics in unforeseen ways is already apparent. It is a sign of things to come that when Bush stepped before the television cameras on August 9 for his first presidential address to the American people, it was not to discuss scrapping the ABM Treaty for missile defense, or the country's tense relations with China, or reform of Social Security, or his tax cut—all of which had been prominently debated during the 2000 election. Instead, it was to explain his decision to allow federal funding for limited research on embryonic stem cells—cells that no one knew how to isolate in a lab until as recently as three years ago and that had

This essay was originally published in *The Public Interest*, Winter 2002.

gone virtually unremarked upon during the presidential election just nine months earlier. Yet the controversy surrounding these cells had by mid summer reached such a feverish pitch that Bush felt compelled to address the issue in a nationally televised address.

Alexis de Tocqueville knew something about the hazards of predicting the future. As he explained in the introduction to *Democracy in America*, he "wished to consider the whole future" of democracy, and he boldly claimed to have seen further into that future than any of his contemporaries. It was not an idle boast, and even today, over 170 years later, his prescience is remarkable. Somehow a man of the early nineteenth century managed to know us better than we know ourselves. However, Tocqueville was not above miscalling a few big ones. He thought that the problem of slavery would end in a race war and the extermination in the South of either the black or white races. What he could not have possibly foreseen was that an exceptional statesman, Abraham Lincoln, would emerge, and that Lincoln would be followed by many others, black and white, who would dedicate themselves to making multiracial democracy work.

But predicting the exact course of events was never in fact what Tocqueville set out to do in *Democracy in America*. As he explained, in the very chapter in which he failed to see the direction of race relations in the United States:

> It is difficult enough for the human mind to trace some sort of great circle around the future, but within that circle chance plays a part that can never be grasped. In any vision of the future, chance always forms a blind spot which the mind's eye can never penetrate.

Tocqueville's aim was to draw "the great circle" of America's future—that is, to define the possibilities and limits of American democracy. Prophesizing, Nostradamus-like, what might occur within that circle was no part of his business, for here fortune alone rules.

In considering America's future in relation to the biotechnology breakthroughs that have already arrived, and to those that are still to come, we should follow Tocqueville's example. We should worry less about this or that invention and focus instead on the great circle of our future. The challenge, for all of its complexity, can be simply stated: Biotechnology holds out tremendous promise for curing disease but also tempts us with the conquest of human nature in the form of a new eugenics. The question is, what general ideas and opinions in our culture might prevent us from drawing a line between the promise and the peril of biotechnology?

THE THREE WAVES OF MEDICINE

Before tackling this question, a few things should be said about immediate and near-at-hand developments in biotechnology. In surveying these developments, I will depend on Nicholas Wade's recent book, *Life Script: How the Human Genome Discoveries Will Transform Medicine and Enhance Your Health*. Wade is a highly reliable guide. A former writer for *Nature* and *Science* magazines, two preeminent science journals, he is currently a science reporter for the *New York Times*, where he writes on cutting-edge developments in biotechnology.

In Wade's view, the new genomic technologies will "develop in three overlapping waves of innovation which can be called conventional, germline, and life-extending." The conventional therapies will include powerful diagnostic tools known as "gene chips" that will be able to scan a person's genome and detect its vulnerability to various diseases. Meanwhile, by making use of the recently decoded human genome, scientists will soon develop powerful, protein-based wonder drugs. These new diagnostic tests and drugs will mark the dawn of what Wade calls "individualized medicine." Patients will be treated based on their particular genetic makeup, with drugs matched to individual genotypes. Also on the immediate horizon is what Wade calls "regenerative medicine." Instead of merely helping people to live longer with disease, medical science will harness the power of stem cells to grow into any number of replacement parts and organs. When a patient gets liver disease, for example, he will be supplied with a newly grown liver. Such advances in science will, Wade concludes, lead to the fulfillment of conventional medicine's "ultimate goal": good health to the end of one's days, which will add up to, on average, 120 years.

The second wave, Wade speculates, will involve germ-line manipulation of the human genome. In germ-line therapy, new genes are inserted directly into the fertilized egg; the resulting genetic changes are permanent, passed on from one generation to the next. Initially, germ-line therapy might be used to repair clear-cut genetic defects, and few people would object to that use of the technology. But, at least in theory, the technique could also be used for enhancement purposes, such as increasing height and boosting IQ or tinkering with the emotional and behavioral makeup of humans. Already, scientists have reportedly increased intelligence in mice by means of genetic manipulation.

Wade's third wave of medical innovations involves a radical increase in the "natural" life span of the human species. Already, scientists have extended the life span of fruit flies, roundworms, and mice. In the case of the roundworm, life span was quadrupled. As Wade points out, if the same were accomplished in humans, people who now live to the ripe old age of 80 years could reach 320 years.

If modern medicine's march ended with Wade's first wave, the social effects would arguably be minimal. The extension of the life span to 120 years would only complete a trend underway since the beginning of the last century, when average life expectancy was in the mid forties. It is not clear that a jump from 80 to 120 would be any more dramatic than was the jump from 40 to 80. One might ask whether Social Security would remain solvent with people living so long, but that is already a concern. One could raise other, more difficult questions about a nation in which the average age has greatly increased. Would the social ties of affinity and affection between generations remain strong if the "young" must wait until their mid seventies to hold high office or to reach the top of the corporate ladder? Would a nation of senior, senior citizens be willing to risk blood and treasure to defend itself? Would it still launch missions to the moon or send men aloft into space? Would a nation of 90-year-olds still have the capacity to consider questions anew? Or are spiritedness and a sense of wonder qualities mainly of the young? These are serious but speculative matters, which are unlikely to lead people to object to medicine's advance. Only antiscience zealots and a few exceptional stoics would say no to a long and healthy life for themselves and their children. (Of course, objections of a different sort have been raised about the morality of the *means* employed in the course of this "first wave," means that may involve experimenting on human embryos.)

In contrast, the second and third waves will indeed raise profound and unprecedented questions. Edward O. Wilson, the highly regarded Harvard sociobiologist, has predicted that within the next several decades, an era of "volitional evolution" will commence, in which it will be possible not only to increase intelligence and other such qualities but also to transform the basic emotional drives of *Homo sapiens*. One does not have to be religious to question whether man has the wisdom to gamble with human nature in this manner. Wilson himself asks why our species would "give up the defining core of its existence, built by millions of years of biological trial and error." Not only do we lack the knowledge to tinker with human nature, but there is something immoral and tyrannical about one generation of human beings experimenting upon and changing the essence of humanity for all succeeding generations.

There are other, darker possibilities, human cloning among them. Despite the fact that animal clones suffer from high rates of deformity, and despite the many ethical objections to human cloning, several fertility experts in America and abroad have announced their intention to clone a human being. There is as of this writing no legal barrier standing in their way. The creation of human chimeras is another horrifying possibility. Several years ago, a biotechnology company claimed to have created an embryonic cell that was part cow, part human, the result of a fusion of a human cell nucleus and a cow egg. The embryo was allowed to grow and divide five times, though there was apparently no intention to transfer it to a uterus. This experiment was not the work of some ghoulish Dr. Kevorkian but of well-respected scientists at the established biotechnology company Advanced Cell Technology in Massachusetts. As the *New York Times* reported at the time, "A perplexing feature of the hybrid embryo would be that it would start mostly bovine, then become mostly yet not entirely human." Perplexing indeed, and yet perfectly legal under current law.

TODAY'S DEBATE

Given the indeterminacy of events, it seems pointless to second guess Wade and other experts in the field. Whether it will indeed become possible to enhance IQs or to quadruple the human life span is anyone's guess, though it is worth noting that these are the predictions of the scientists themselves, not the writers of science fiction. The mind boggles at what life in America would be like once the country is engulfed by Wade's second and third waves, by a freewheeling eugenics of cloning and enhancement. Yet what is within reach of analysis is the question of how our culture will respond to such breakthroughs—that is, whether it will view the new discoveries with skepticism, if not repugnance, or whether it will embrace every new possibility. For here, the area of Tocqueville's great circle, if you will, we need merely examine what is being said now about biotechnology to reach some conclusions about what will be said in the future. Of course, there is no way of knowing which arguments will prevail, for one never knows from what corner an Abraham Lincoln will emerge, transcending and transforming the political debate. But we can make some reasonable surmises nonetheless.

Today's culture wars over biotechnology concern two issues, whether to pass legislation that would criminalize reproductive as well as therapeutic cloning and whether stem cell research on embryos should receive the financial backing of the federal government. In the case of cloning, the House of Representatives has passed by a vote of 265 to 162 the Weldon bill, which would criminalize cloning of human embryos for any purpose, reproductive or therapeutic. What kind of anti-cloning law will eventually, if ever, be passed, now depends upon the Senate, where the Democrats hold a slim majority. As for federal support for research on stem cells, Bush settled on a compromise that allows the federal government to fund existing stem cell lines but not the creation of any additional ones.

The cloning and stem cell debates produced an unusual political alliance between religious conservatives and the environmental Left, both of whom favored either government regulation of, or outright bans on, the new techniques. They are opposed by liberal Democrats and the biotechnology industry. This political debate is very important and bears close watching in the years ahead. But deeper cultural trends are also at work, to which one might apply Tocqueville's label of "mores"—that is, "the habits of the heart," "the different notions possessed by men, the various opinions current among them, and the sum of ideas that shape mental habits." Of these habits, notions, opinions, and ideas, three stand out: nonjudgmentalism, scientism, and egalitarianism. Each of these, which I will discuss in turn, tends to favor the unchecked advance of the biotechnology revolution.

NONJUDGMENTALISM

Whether it gains sustenance from relativism, postmodernism, or libertarianism, nonjudgmentalism has become one of the distinguishing marks of our culture. In most matters, Americans think it is wrong to judge others or to attempt to constrain their choices. The public's nonjudgmentalism was on display most recently during the Clinton scandals. According to opinion polls, the majority of Americans thought that Clinton was wrong to have had a sexual affair with a young intern and then lie about it, but they also believed that it was his private concern and none of their business. What happens when such nonjudgmentalism meets biotechnology? Will cloning and genetic enhancement be viewed as the private affairs of individuals?

There is little polling data available on what people think about future biotechnology developments, with the possible exception of cloning. On the question of human or reproductive cloning, Americans, at least for now, tend to be quite judgmental. Generally, upwards of 90 percent of the public opposes the cloning of human beings, though this opposition slackens somewhat among the young. A 1997 poll sponsored by *Time* and CNN found that 64 percent of 18-to-29-year-olds favored therapeutic cloning—that is, cloning for replacement tissues and parts—compared to only 32 percent of those over 65 years.

Yet it is not enough to look at polling data. If opinion polls were the decisive measure of public sentiment, why has Congress thus far failed to ban reproductive cloning? Given the lack of legal action, one might question the depth of Americans' opposition. As many pundits have noted, when the first baby clone is born, with its cute face splashed on the cover of every magazine, opposition is likely to melt away. In other words, which laws get passed and which laws do not is at least as important a measure of public sentiment as opinion polls.

The popular culture can also tell us something about public opinion on these issues. Consider a recent Hollywood film about cloning, starring Arnold Schwarzenegger. The movie, entitled *Sixth Day*, takes place in the near future: Human cloning has been made illegal after some disastrous experiments, but corporate America wants the ban overturned, a ban that is strongly supported by a motley crew of religious zealots. Arnold plays a typical American who, though not religious, has qualms about cloning. When his daughter's pet dog dies, he will not pay for a clone-replacement because he finds the idea simply too weird. His anticloning prejudice seems confirmed when an evil corporate CEO mistakenly clones him, unleashing all sorts of mayhem. But this is no update of the Frankenstein story, warning of the hubris of science, for Arnold comes to appreciate that his clone is a person too, with human feelings and a soul. And even though his clone has sex with his wife, Arnold befriends him anyway. It's only a matter of loving himself, and what could be more natural and pleasing to democratic taste than self-love? By movie's end, Arnold, now enlightened, clones his daughter's dead pooch, and everyone, he and his clone, live happily ever after.

More sophisticated versions of nonjudgmentalism spring from the ideologies of postmodernism and libertarianism. Just as biotechnology has led religious conservatives to work with environmental Greens, so too it has led conservative libertarians to join hands with the postmodern Left.

It is, of course, no secret that over the last several decades, postmodern ideas have become increasingly prevalent, most notably in academic circles. From a postmodern perspective, developments in biotechnology might seem both to confirm that "human nature" is entirely plastic and to open the door to new, experimental modes of being. Harvard law professor

Laurence Tribe's writings on these issues illustrate postmodernism's influence on the biotechnology debate. In the early 1970s, when America's first great cloning debate occurred (somewhat prematurely, as it turned out), Tribe opposed the cloning of humans on the traditional liberal grounds that it amounted to using humans as means and not ends. Yet 25 years later, after the arrival of Dolly the cloned sheep, Tribe reversed his position. Under the influence of postmodernism, he argued that it cannot be known whether cloning is degrading of human nature, which he now, in good postmodern fashion, puts in quotation marks. Furthermore, he attacked those who opposed cloning as "essentialists" and claimed that a ban on human cloning risks, in his words, "cutting [society] off from vital experimentation and risks sterilizing a significant part of its capacity to grow."

Those influenced by postmodernism do not view cloning as something to be merely tolerated or as the price society must pay for the medical benefits that this technique might yield. Instead, they see cloning as a positive good, a way of breaking down old habits of thought and overturning what they view as harmful prejudices. To them, cloning is the ultimate "transgression."

A more buttoned-down version of nonjudgmentalism can be found on the Right among those who call themselves libertarians. They argue with respect to biotechnology, as they do with most other social questions—whether it is gambling, pornography, drugs, or divorce—that it is none of the government's business what consenting adults do with their time, money, and bodies. A ban on cloning is the "road to serfdom" all over again. In their view, we should trust individuals to know what is best for themselves. Conveniently, libertarians overlook the fact that eugenics involves third parties who have not consented to the proposed genetic engineering, whether it is the clone-to-be, the "designer baby," or the new race of "posthumans."

How far nonjudgmentalism might eventually reach can be glimpsed from the *New York Times* article that I mentioned earlier. The reporter acknowledged that the American public would be greatly troubled by the possibility of mixing man with animal, but he added the key qualifier, "at least at first." For as it happens, many scientists and bioethicists already believe that this is a rather outdated distinction. Said one expert: " 'Biologically a lot of this research is showing us similarities and the upshot in a hundred years may be that the lines between humans and nonhumans will be viewed as a little bit grayer.' " A slogan for the next century might be: "Down with speciesism!"

SCIENTISM

Nonjudgmentalism finds an ally in what I will call "scientism," which may ultimately pose a more fundamental, and perhaps less familiar, challenge. Today, many believe that scientists should be the judge of their own cause, free of the meddlesome control of moralists, philosophers, theologians, and politicians. Examples of this mindset are not hard to find. In an article on Bush's stem cell deliberations, *New York Times* columnist Frank Rich complained that "political science [is taking] precedence over biological science." Similarly, after Bush's stem cell decision, Representative Richard Gephardt objected that "Americans know this is not the decision that the science community needs to go full force."

Note that word "force," for it is a favorite of those, whether liberal or conservative, who are under the sway of scientism. Earlier in the year, for example, former Speaker of the House Newt Gingrich made a fetish of force when he urged his Republican colleagues to replace their conservative principles with a simple belief in, as he put it, "technology as a force of change." A variation on this theme holds that scientific "progress" is inevitable, and so the attempt to stop it is futile. There is a certain irony in this notion of science as unstoppable force. Modern science, whose overriding goal is the conquest of the blind forces of nature, is now itself considered a blind force beyond human control.

As long as medical science remained dedicated to relief of suffering, no democratic oversight was required. For relief of suffering is, almost by definition, *the* democratic goal. Yet today, medical science is on the verge of finding a new polestar—one other than health—by which to organize its activities and to guide its progress. This is the crucial thesis of Nicholas Wade's book. In his words:

> Though [today's genomic] revolution is being conducted in the name of medicine, it will not necessarily stop at its implied goal, the attainment of perfect health. The power to reshape the human clay has no clear limits. How far should we go in enhancing qualities other than health, such as physique or intelligence?

It is a question that haunts Wade. "Increasing life expectancy lies fully within the agreed goal of contemporary medicine," he states, while "increasing the life span is a different and far more ambitious goal." Once "the ultimate goal of conventional medicine," namely health, is achieved, "radical departures" like germ-line engineering are in the bidding, he writes. "The true dangers of genome engineering," he warns, "lie in the question of what changes should be permitted, if any, other than those directly related to health."

In writing that the "implied," "agreed," and "ultimate goal" of medicine is health, Wade is trading on an understanding of the modern scientific project formulated four hundred years ago by Francis Bacon. On the one hand, Bacon wished to liberate the pursuit of knowledge from the shackles of religion and the stultifying influence of ancient philosophy. He foresaw the kind of science we have today, one that produces all sorts of powerful inventions and cures. But on the other hand, in order to prevent the misuse of this liberated science, Bacon proposed that it be given one exclusive and binding directive: "the relief of man's estate." In *The Great Instauration* he starkly warned:

> Lastly, I would address one general admonition to all; that they consider what are the true ends of knowledge, and that they seek it . . . but for the benefit and use of life; and that they perfect and govern it in charity.

As several scholars have noted, Bacon's scientist was to be, in a sense, a secular priest. He would devote himself to the conquest of nature, but only for the altruistic purpose of ministering to the sick. He would be full of curiosity about the natural world and ambition to subdue it, but he would also be full of charity for the lot of man.

In his 1966 book *The Phenomenon of Life*, the philosopher Hans Jonas scrutinized charity's role in directing and giving purpose to modern science, and argued that Bacon's attempt to graft charity onto science was bound to fail eventually. The problem is that the scientist, strictly in his capacity as a scientist, cannot justify such a goal, for the intellectual and moral sources of charity are independent of the scientist's discipline. As Jonas formulated the quandary:

> The need for charity or benevolence in the use of theory stems from the fact that power can be for evil as well as for good. Now, charity is not itself among the fruits of theory in the modern sense. As a qualifying condition of its use—which use theory itself does not specify, let alone assure—it must spring from a source transcendent to the knowledge that the theory supplies.

From Bacon's time to our own, medical science has by and large limited itself to the charitable aim of making imperfect men comfortable. But now, it is clear, many scientists have set their sights on the new goal of perfecting imperfect men, not to mention such oddities as cloning. And in today's secularized culture, it is increasingly difficult to insist, as Bacon once did, that scientists should be governed by a religious or ethical norm like charity. Such a norm is dismissed as the residue of a faded religious faith, a faith that must not be forced on nonbelievers, or a humanism that has been discredited. Meanwhile, others, influenced by the nonjudgmentalism of the day, think it none of their business to dictate goals and values to scientists. So what if cloning or genetic enhancement are largely unrelated to the relief of suffering; why, it is demanded, should this remain science's goal, especially today when science is capable of doing so much more? E. O. Wilson considers this "the ultimate question: To what end, or ends, if any in particular, should human genius direct itself?" Bacon's answer evidently no longer suffices.

EQUALITY

Nonjudgmentalism and scientism would prepare the way for a laissez-faire eugenics. People would choose what genetic enhancements they please and can pay for, and scientists would create as they like. But there is another widely held notion in America—egalitarianism—which pulls in a different direction. It too will make resistance to the biotechnology revolution difficult, but the outcome will be instead of a laissez-faire eugenics some form of government-subsidized and regulated eugenics.

In his 1997 book *Remaking Eden,* Princeton molecular biologist Lee Silver speculates that in the future two genetic classes will emerge, the "GenRich," a hereditary class of genetic aristocrats, who take advantage of the new science of eugenics, and the "Naturals," who cannot afford to enhance their offspring. Whatever Silver's competency in science, his political analysis is probably off the mark. As Francis Fukuyama has pointed out, if genetic enhancement were to become possible, democratic publics would take to the streets with knives and guns before allowing Silver's scenario to come to pass. The lower and middle classes would insist that their children be provided with the same eugenic enhancements available to the children of the rich. In time, the U.S. government would subsidize eugenic programs, not to create an overclass but to preserve equality, to elevate everyone's natural endowments.

Indeed, some of today's most influential liberal thinkers have defended eugenics on these grounds. That they should do so is especially interesting, if not surprising, since liberals have for the better part of the last 50 years attacked genetic explanations of human behavior as discredited holdovers of Nazism and fascism. At the same time, however, liberals have also denied the possibility of "equality of opportunity." Individuals, they point out, each begin life from a particular socio-economic class and with different natural abilities. This realism, as it were, about the limits of equality in a free society ultimately softened liberal objections to eugenics.

In John Rawls's classic *A Theory of Justice*, published in 1971, long before the biotechnology revolution, a curious passage appeared. Rawls states that he will not consider the question of eugenics but then notes briefly that it is

> in the interest of each to have greater natural assets. This enables him to pursue a preferred plan of life. In the original position, then, the parties want to insure for their descendants the best genetic endowment (assuming their own to be fixed). The pursuit of reasonable policies in this regard is something that earlier generations owe to later ones, this being a question that arises between generations. Thus over time a society is to take steps at least to preserve the general level of natural abilities and to prevent the diffusion of serious defects.

Rawls concludes by stating, "I shall not pursue this thought further," but thirty years later, with the potential of genetic engineering more evident, his disciple Ronald Dworkin is not so reticent. In his 2000 book, *Sovereign Virtue: The Theory and Practice of Equality*, he vigorously and at length defends genetic engineering on the grounds of equality. The equality principle requires, in his view, that we do all in our power—including eugenics—to make each life a successful one. The step from government redistribution of income in the name of equality to government-sponsored eugenics in the name of equality is apparently a small one.

At the end of his book, Nicholas Wade captures the strange marriage that might come about between the democratic passion for equality and the science of eugenics. "The sequencing of the human genome makes it possible to envisage for the first time the creation of a genetically more just society, one in which the most fundamental kind of wealth—the genes that confer health and fitness—would for the first time be accessible to all."

LOOKING TOWARD THE FUTURE

Nonjudgmentalism, scientism, and equality—these three notions form the backdrop against which the biotechnology revolution is unfolding. Together, they make up, as Tocqueville would say, the "great circle" of our future. This does not mean that it is futile to resist biotechnology's advance. I do not wish to spread despair. Congress will in all likelihood enact a ban against human cloning, and President Bush has made clear that the federal government will not unconditionally endorse experimentation on human embryos. The president has also appointed the bioethicist Leon Kass to head a national commission with the broad mandate of guiding policy makers and the public through the moral minefield of the biotechnology revolution. These are important counter-developments to the trends I have canvassed.

Then there is the role that chance plays. The attack on the Pentagon and the World Trade Center could transform our culture in deep and lasting ways. Nonjudgmentalism may not seem quite so laudable in this other brave new world we have now unfortunately entered—a world of terror on our very own shores. And bioterrorism in the form of anthrax and smallpox reminds us, more vividly perhaps than any theoretical argument, that the life sciences can also be used for evil and death. Though hardly worth the price, we may regain our moral compass and rediscover what has always been most admirable about the American experiment: Not so much our science and technology, as wonderful and inspiring as these are, but our democratic way of life, our example to the world of a free people governing itself.

If nothing else, the events of September 11 remind us not to become overly distracted by the biotechnology revolution. It raises many profound questions but poses no immediate crisis. Criticism of biotechnology's progression must not be allowed to develop into something else, an indictment of America in its entirety. For whatever the country's excesses, and there are no doubt many, America remains the last, best hope of good-willed people everywhere.

OF TERRORISM AND CLONING

Eric Cohen and William Kristol

In testimony before the Senate last July, Dr. Michael West, president of Advanced Cell Technology and the lead scientist on the team that recently cloned the first human embryos, scolded legislators for thinking like children. To make his point, he quoted the New Testament:

> As the Apostle Paul said: "When I was a child, I spake as a child, I understood as a child, I thought as a child: but when I became a man, I put away childish things." (I Cor 13:11) In the same way it is absolutely a matter of life and death that policy makers in the United States carefully study the facts of human embryology and stem cells. A child's understanding of human reproduction simply will not suffice and such ignorance could lead to disastrous consequences.

A few months later, Dr. West made the historic announcement that his company had successfully cloned human embryos in the hope of one day harvesting their stem cells. When asked whether he believed his technique would eventually be used to clone newborn human beings, he replied: "I'm not an expert in ethics. . . . But, biologically, scientifically, I don't know of any reason why that would not happen."

Perhaps Dr. West needs a reminder of his own biblical principle, this time applied not to scientific facts but to the moral obligations of scientists: A child's understanding of ethics will not suffice; such ignorance could lead to disastrous consequences.

But in reality, Dr. West does not believe his research is morally neutral. He is not a nihilist. Just the opposite: He believes that healing the sick is the highest human good, and therefore the only necessary justification for harvesting human clones. "For the sake of medicine," he informs us, "we need to set our fears aside." For the sake of health, we need to overcome our moral inhibitions against cloning and eugenics.

For the first half of 2001, the prospect of a brave new world—and specifically, the issues of human cloning and embryonic stem cells—had come to center stage in American politics.

Then came the terrorist attacks of September 11. Fears of a brave new world were no longer at the forefront of the national mind. It was now death and destruction we feared, not utopian biology. It was bioterrorism we feared, not morally compromising advances in biomedical research. It was human mortality we feared, not a post-human future of would-be immortals.

Perhaps it is significant that the genetic age and the age of terrorism seem to have arrived together. For both require us to confront fundamental questions about life and death, good and evil, civilization and barbarism. The new genetics leads us to expect an indefinite extension of life, to believe that medical science may one day smooth the jagged edges of our mortality. Terrorism confronts us with the permanent fragility of life, and with the destruction modern technology, in the hands of evildoers, may one day unleash upon its creators.

Aldous Huxley understood the connection. In his novel, the brave new world comes into being in large measure as a remedy for human fear—a way of "perfecting" existence so that men and women could lead long, healthy, and pleasure-filled lives. It was an escape from the burdens of history, suffering, and war. As Mustapha Mond explains in *Brave New World*, "What's the point of truth or beauty or knowledge when the anthrax bombs are popping all around you? . . . People were ready to have their appetites controlled then. Anything for a quiet life."

For the last decade, Americans have had a generally quiet life—happy, healthy, upwardly mobile, unburdened by history. The holiday ended when the first plane hit the first World Trade Center tower. What confronts us now is a band of nihilistic terrorists who despise mere health, comfort, and life. Our enemies worship death—not just our death, but their own apocalyptic, civilization-destroying suicide. Osama Bin Laden put it bluntly: "We love death. The U.S. loves life. That is the big difference between us." The challenge to America—a nation that "loves life," and rightfully so—is that confronting such death-seeking terrorism requires a willingness to fight, and perhaps to die. It requires courage, and even heroism.

But even as America prepares for the long fight against terrorism, the nation cannot overlook the moral challenges within, especially the "ethical minefields" created by our own genetic and biological "advances." Hatred of life and glorification of death lead in obvious ways to evil. But an absolute devotion to health and material well-being may invite civilized nations to tolerate, even celebrate, morally questionable pursuits (like cloning human embryos for research or harvesting organs) and morally debilitating expectations (like a life without challenges, tragedy, or suffering). As the philosopher Paul Ramsey put it, "any person, or any society or age, expecting ultimate success where ultimate success is not to be reached, is peculiarly apt to devise extreme and morally illegitimate means for getting there." And peculiarly apt, one might add, to redefine their project so that it seems morally blameless, as Sen. Arlen Specter and other zealous advocates for unlimited research have tried to do, by saying that cloned embryos are not really cloned embryos.

The cloning debate is not simply the latest act in the moral divide over abortion. It is the "opening skirmish," as Leon Kass puts it, in deciding whether we wish to "put human nature itself on the operating table, ready for alteration, enhancement, and wholesale redesign." Lured by the seductive promise of medical science to "end" suffering and disease, we risk not seeing the dark side of the eugenic project: that designing our descendents, whether through cloning or germ-line engineering, is a form of tyranny by one generation over the next, however well-meaning the designers; that in trying to live indefinitely like gods, we end up mixing our genes with those of cows, mice, and jellyfish; that modern man's mistreatment of the natural environment demonstrates his inability to see the unintended consequences of his powers; and that in trying to stamp out disease by any means necessary, we risk beginning the "compassionate" project of stamping out the diseased themselves (something that has already begun with the selective abortion of "undesirable" embryos).

Taken together, what America now faces are two grave threats to a dignified human future—one which is obvious, and one which comes so wrapped up with real and apparent goods that it is hard to detect. The first is the dehumanization of the terrorists, who have so little regard for life (including their own) that they make killing their only purpose and modern technology their weapon. The second is the dehumanization of the brave new world, which will come, if it does, from the false faith that science and technology can make human (or post-human) life perfectly healthy, pleasant, autonomous, and secure, even if a few "eggs" may need to be cracked along the way.

Standing up to these twin threats requires a similar courage, virtue, and moral realism. In a recent essay on the moral parallels between war and medicine, Gilbert Meilaender quotes the stoic philosopher Marcus Aurelius: "Another [prays] thus: How shall I not lose my little son? Thou thus: How shall I not be afraid to lose him." In other words, if my son must fight and even die to face and destroy the threats posed by men like Hitler or Bin Laden, I pray I have the courage to let him do so. And if he must die because the only way to save him would be for me to do evil, I pray I have the courage to let him die rather than destroy the principles that give his life value in the first place. These are not easy questions or easy choices. They are not pleasant subjects to think about. But they are no less real for being unpleasant.

Perhaps the clearest way to think about these moral questions is to consider what kind of world we will have if we act and what kind if we don't. On the one hand, a nation that stands idle while terrorists plot its destruction has little chance of surviving. And a great nation that stands idle while tyrants slaughter their own people has little claim to greatness. On the other hand, a nation that begins the project of putting "death to death" by any means necessary—even if some genuine good might come from it—devalues life in the misguided effort to extend it indefinitely. It risks undermining the most sacred medical principle of "do no harm," and risks leading us (whether enthusiastically or unawares) toward what C. S. Lewis saw long ago as "the abolition of man." As Lewis put it: "Each new power won *by* man is a power *over* man as well. Each advance leaves him weaker as well as stronger. In every victory, besides being the general who triumphs, he is also the prisoner who follows the triumphal car."

It may be that after September 11 the Baconian project of "relieving man's estate" will seem more urgent than ever, now that health and security seem so precarious. But it is at least worth noting that it is modern technology itself that has made our own technological response necessary: new weapons require new counter-weapons, new missile technologies require new anti-missile technologies, new chemicals of death require new chemical medicines. In the end, the tools of extinction and survival are often indistinguishable. Progress and destruction can spring from the same source.

Hans Jonas, the great philosopher of science, understood this dilemma:

> Modern technology, informed by an ever-deeper penetration of nature and propelled by the forces of market and politics, has enhanced human power beyond anything known or even dreamed of before. It is a power over matter, over life on earth, and over man himself; and it keeps growing at an accelerating pace. Its unfettered exercise for about two centuries now has raised the material estate of its wielders and main beneficiaries, the industrial "West," to

> heights equally unknown in the history of mankind. Not even the ravages of two world wars—themselves children of that overbrimming power—could slow the upward surge for long: it even gained from the spin-off of the hectic technological war effort in its aftermath.

"But lately," Jonas continued,

> the other side of the triumphal advance has begun to show its face, disturbing the euphoria of success with threats that are as novel as its welcomed fruits. Not counting the insanity of a sudden, suicidal atomic holocaust, which sane fear can avoid with relative ease, it is the slow, long-term, cumulative—a peaceful and constructive use of worldwide technological power, a use in which all of us collaborate as captive beneficiaries through rising production, consumption, and sheer population growth—that poses threats much harder to counter. The net total of these threats is the overtaxing of nature, environmental and (perhaps) human as well. Thresholds may be reached in one direction or another, points of no return, where processes initiated by us will run away from us on their own momentum—and toward disaster.

To bow to such inevitable disaster—either in the form of a "sudden atomic holocaust" or the form of "peaceful biological advances"—is to give up on the core of the American experiment: "that honorable determination," as James Madison put it in *Federalist No. 39*, "to rest all our political experiments on the capacity of mankind for self-government." In other words, men and women ought to resist becoming either slaves or slaveholders. Every people and every generation should have the freedom to govern themselves by reflection and choice—but to do so without the hubristic belief that self-government means perfect government, or that a nation of self-made men means creating a post-human civilization of man-made selves. This moral realism and freedom is what America stands for, and it is what, in very different ways, the terrorists and the eugenicists both seek to deny.

INDEX

ABOUT THE EDITORS

William Kristol is editor of *The Weekly Standard* and one of the nation's leading political analysts and commentators. He served as chief of staff to Vice President Dan Quayle during the Bush administration and to Secretary of Education William Bennett under President Reagan. He taught politics at the University of Pennsylvania and Harvard, and co-edited *The Neoconservative Imagination: Essays in Honor of Irving Kristol* and *Educating the Prince: Essays in Honor of Harvey Mansfield*. He also serves as chairman of The Bioethics Project.

Eric Cohen is a resident scholar at the Ethics and Public Policy Center. His essays and articles have appeared in *The Weekly Standard*, *The Wall Street Journal*, *The Los Angeles Times*, *First Things*, and many other magazines and newspapers. He was formerly a fellow at the New America Foundation and managing editor of *The Public Interest*. He also serves as a consultant to the President's Council on Bioethics.